WORKBOOK TO ACCOMPANY

Automotive Maintenance & Light Repair

Second Edition

Rob Thompson

CENGAGE

Australia • Brazil • Canada • Mexico • Singapore • United Kingdom • United States

Workbook to Accompany Automotive Maintenance & Light Repair
Second Edition
Rob Thompson

SVP, GM Skills & Global Product Management: Jonathan Lau

Product Director: Matthew Seeley

Senior Product Manager: Katie McGuire

Senior Director, Development: Marah Bellegarde

Senior Product Development Manager: Larry Main

Senior Content Developer: Meaghan Tomaso

Product Assistant: Mara Ciacelli

Vice President, Marketing Services: Jennifer Ann Baker

Associate Marketing Manager: Andrew Ouimet

Senior Content Project Manager: Kenneth McGrath

Design Director: Jack Pendleton

Text Designer: Jay Purcell

Cover Designer: Dave Gink

Cover Credit: Boophuket/Shutterstock.com

For product information and technology assistance, contact us at **Cengage Customer & Sales Support, 1-800-354-9706**

For permission to use material from this text or product, submit all requests online at **www.cengage.com/permissions.** Further permissions questions can be e-mailed to **permissionrequest@cengage.com**

Library of Congress Control Number: 2017952177

ISBN: 978-1-337-56440-3

Cengage
200 Pier 4 Boulevard
Boston, MA 02210
USA

Cengage is a leading provider of customized learning solutions with employees residing in nearly 40 different countries and sales in more than 125 countries around the world. Find your local representative at **www.cengage.com.**

To learn more about Cengage Learning, visit **www.cengage.com**

Purchase any of our products at your local college store or at our preferred online store **www.cengage.com**

Notice to the Reader
Publisher does not warrant or guarantee any of the products described herein or perform any independent analysis in connection with any of the product information contained herein. Publisher does not assume, and expressly disclaims, any obligation to obtain and include information other than that provided to it by the manufacturer. The reader is expressly warned to consider and adopt all safety precautions that might be indicated by the activities described herein and to avoid all potential hazards. By following the instructions contained herein, the reader willingly assumes all risks in connection with such instructions. The publisher makes no representations or warranties of any kind, including but not limited to, the warranties of fitness for particular purpose or merchantability, nor are any such representations implied with respect to the material set forth herein, and the publisher takes no responsibility with respect to such material. The publisher shall not be liable for any special, consequential, or exemplary damages resulting, in whole or part, from the readers' use of, or reliance upon, this material.

Printed in the United States of America
Print Number: 05   Print Year: 2021

Contents

Preface

The *Workbook to Accompany Automotive Maintenance & Light Repair*, 2nd Edition is intended to reinforce comprehension of chapter content from the core textbook, to support critical thinking about the material learned, and finally, to allow students to put their knowledge to practice in the shop. Each Workbook chapter includes a section of Review Questions to allow additional review, reflection, and assessment on the chapter content beyond the chapter-end questions in the textbook. Workbook Activities are designed to further ingrain the material in interesting and meaningful ways, such as part/component identification exercises and in-class experiments. Finally, Lab Worksheets offer step-by-step, guided instruction through the types of basic inspection, testing, and maintenance procedures that an entry-level technician is likely to perform on the job.

CHAPTER 1

Introduction to the Automotive Industry

Review Questions

1. With the introduction of _____, major changes started to take place in the automotive industry.

2. Currently, about _____ of all jobs in the United States are directly related to transportation.

3. Which of the following has been a factor in modern automobile design?
   a. Emissions
   b. Safety
   c. Fuel economy
   d. All of the above

4. An _____ technician is expected to be able to perform basic inspections and maintenance services.

5. List six services and repairs often performed by entry-level technicians.
   a. _____
   b. _____
   c. _____
   d. _____
   e. _____
   f. _____

6. As important as technical skills are, it is also important to be able to _____ and _____ technical information.

7. Which of the following are likely to be the most experienced technicians in the shop?
   a. A-level technicians
   b. B-level technicians
   c. C-level technicians
   d. None of the above

8. Technicians who work in specialty shops that repair and rebuild engine components are called automotive _____.

9. A _____ technician performs body and paint repairs.

10. A _____ _____ communicates with the customer and routes the service order to the technicians.

11. Ford ASSET, Toyota T-Ten, and BMW STEP are manufacturer-sponsored training programs available at _____ schools.

12. Describe the difference between a degree and a diploma program.
    _____
    _____

13. The _____ _____ _____ _____ _____ is a partner with ASE and accredits automotive training programs.

14. Continuing to learn and acquire new skills is called _____ learning.

15. ASE offers _____ general automotive certifications.
    a. four
    b. six
    c. eight
    d. nine

16. Once a person passes ASE tests A1 through A8, he or she is considered an ASE _____ Automobile Technician.

17. ASE Student Certifications are valid for _____ years.

Activities

1. Using an Occupational Outlook Handbook or the OOH sections of the Bureau of Labor Statistics website, research the following information about automotive careers.

   a. Describe skills necessary to be successful as an automotive technician.

   _____

   _____

   b. List some of the services a technician is likely to perform in his or her job.

   _____

   _____

   c. Describe the automotive work environment.

   _____

   _____

   d. Describe the educational options available to further your education in auto technology.

   _____

   _____

   e. What must a person do to become ASE certified?

   _____

   _____

   f. Describe the job outlook in the automotive repair industry.

   _____

   _____

   g. How many people are employed as automotive technicians in the United States?

   _____

   h. What is the salary range for auto technicians?

   _____

   _____

2. Answer the following questions about your school and the automotive program.

   a. Is the school secondary (high school) or postsecondary? _____

   b. If yours is a secondary school, with what postsecondary school does your program partner for continuing education? _____

   c. How long in semesters, months, or years does it take to complete the auto tech program?

   _____

    d. What are the ASE/ASE Education Foundation subject areas covered by the auto program?

_____

_____

    e. How many of the students obtain jobs in the auto industry during or after completing the auto technology program each year? _____

3. Using the ASE website, www.ase.com, explore ASE certification to complete the following section.

    a. Provide a brief history of ASE and explain how to become ASE certified.

_____

_____

_____

_____

    b. How long is the work experience requirement to become a certified technician?

_____

_____

    c. What are the nine basic automotive certifications?

_____

_____

_____

_____

    d. Describe how test questions are developed.

_____

_____

    e. How and when are ASE tests offered?

_____

_____

Lab Activity 1-1

Name _____ Date _____ Instructor _____

Year _____ Make _____ Model _____

Engine _____ VIN _____

Identify safety and technology systems on a late-model vehicle

1. Determine the number and location of air bags on this vehicle.

   _____

   _____

   _____

2. List other types of safety technology installed on the vehicle. This includes accessories such as backup cameras, rear collision detection, blind spot detection, and others.

   _____

   _____

3. List the electronic accessories found on the vehicle. This includes navigation, DVD entertainment, Bluetooth, music integration, and other items.

   _____

   _____

4. List passenger comfort accessories found on the vehicle. This includes power seats, heated and cooled seats, dual-zone climate control, and other items.

   _____

   _____

5. Using the vehicle information and the website www.fueleconomy.gov, locate the estimated fuel mileage for this vehicle._____

   If the vehicle you are using is five years old or older, try to find either the same make and model or a similar vehicle that is five years newer and compare the fuel mileage ratings.

   How do the two vehicles compare? _____

   If there is a difference, why do you think the difference exists?

   _____

   _____

CHAPTER 2

Safety

Review Questions

1. No aspect of automotive repair and service is more critical than _____.

2. PPE stands for:
   a. Proper protection and equipment
   b. Proper protective equipment
   c. Personnel protective equipment
   d. Personal protective equipment

3. Shop eyewear may consist of which of the following:
   a. Full face shields
   b. Safety glasses
   c. Welding helmets
   d. All of the above

4. Safety glasses should have the ANSI _____ rating stamped on them.

5. Mechanic's work gloves can protect a technician's hands against all of the following except:
   a. Cuts and scrapes
   b. Mild burns
   c. Blood-borne pathogens
   d. All of the above

6. Losing your _____ is often a gradual process and may be less noticeable than other types of injuries.

7. Shop footwear typically requires boots or shoes that are:
   _____

   _____

8. Describe the type of footwear required by your school.
   _____

7

9. A _____ should be worn when working with brake, clutch, or other airborne dust or chemicals.

10. Maintaining good personal _____ means keeping your hair washed and combed as well as taking frequent showers and washing your work clothes.

11. Define *work ethic*. Describe what having a good work ethic means to you._____
    _____
    _____

12. Make a list of six safe work habits.

    a. _____
    b. _____
    c. _____
    d. _____
    e. _____
    f. _____

13. What is the purpose of marked safety zones in the shop?

    _____
    _____

14. Never work under a vehicle supported only on a _____ _____ because failure of the hydraulic cylinder will allow the vehicle to drop, causing injury or death.

15. All lifts have a mechanical _____ that should apply as the lift is raised and automatically engages as the lift is lowered.

16. When moving an engine on an engine hoist, lower the engine close to the floor to lower the center _____ of and reduce the chances of the hoist _____ over.

17. What is meant by dressing a tool?

    _____
    _____

18. Before using a bench grinder, what three items should be checked?

    a. _____
    b. _____
    c. _____

19. Explain why blow guns should never be used while pointing at yourself or another person?

    _____
    _____

20. List five safety precautions for working with air tools.

   a. _____

   b. _____

   c. _____

   d. _____

   e. _____
   _____

21. Describe the safety precautions for using a creeper.

   _____

   _____

22. Describe the typical first-aid procedures for minor cuts and scrapes.

   _____

   _____

23. Explain the differences between first-, second-, and third-degree burns.

   _____

   _____

   _____

24. A _____ can be any substance that can impact public health and damage the environment.

25. List the four identifiers that classify hazardous wastes.

   a. _____

   b. _____

   c. _____

   d. _____

26. Explain four types of information that are contained in an SDS/MSDS.

   a. _____

   b. _____

   c. _____

   d. _____

27. Explain the responsibilities of the EPA.

   _____

   _____

28. OSHA is responsible for overseeing the _____ and _____ of workers.

29. When the battery charges, it releases hydrogen, which can, if exposed to a spark or flame, cause an

   _____.

30. List the correct order in which to disconnect and reconnect battery cables.

_____

_____

31. The first step when charging a battery is to ensure that the battery charger is _____.

32. When jump-starting a vehicle, which connection should be made last?

    a. Dead vehicle battery positive

    b. Dead vehicle engine ground

    c. Dead vehicle battery negative

    d. Good vehicle battery negative

33. The high-voltage wiring and connections on a hybrid vehicle are _____ in color for easy recognition.

34. Describe three precautions to observe before attempting to start a vehicle.

    a. _____

    b. _____

    c. _____

35. The exhaust gas that causes illness and death if inhaled in sufficient quantity is:

    a. Oxygen

    b. Carbon dioxide

    c. Oxides of nitrogen

    d. Carbon monoxide

36. Before moving any vehicle, first perform a _____ to ensure the brake system works properly.

37. The by-products from _____, if allowed to come into contact with your skin, cause skin irritation, rashes, and other health concerns.

38. Many aerosols used in the auto shop are _____ and can easily catch fire if used incorrectly.

39. Waste oil and antifreeze should never be allowed to _____ and must be stored in _____ containers.

40. Brake and clutch friction components may contain _____, a compound that with prolonged exposure, causes lung cancer.

41. A type ABC fire extinguisher contains which of the following?

    a. Water

    b. Dry chemicals

    c. Carbon dioxide

    d. All of the above

Activities

1. Develop a list of personal protective equipment necessary for your class.

_____

_____

_____

_____

2. Match the correct type of PPE with the list of potential shop hazards.

   Mechanic's gloves   Solvent tank or similar chemicals

   Nitrile gloves   Bench grinder

   Chemical gloves   Drum brake service

   Safety glasses   Lifting a cylinder head

   Safety boots   Air hammer

   Ear protection   Blood

   Respirator   Mess on floor

   Back brace   Sharp metal

3. Briefly describe how each of the following shop items presents a danger if used improperly.

   Floor jack _____

   _____

   Vehicle hoist _____

   _____

   Engine hoist _____

   _____

   Bench grinder _____

   _____

   Blow gun _____

   _____

   Creeper _____

   _____

   Battery charger _____

   _____

Impact gun _____

_____

4. On a separate piece of paper, make a rough sketch of your lab and include the following safety items:

First aid kit	Emergency shower	Eyewash station
Fire blanket	Power shut-off	SDS/MSDS
Fire exit	Fire alarm	Fire extinguishers

Identify the posted evacuation routes.

Lab Activity 2-1

Name _____ Date _____ Instructor _____

Locate and inspect jacks and jack stands.

Locate the lab's floor jacks and jack stands and note any safety concerns.

Floor jack caster operation _____

Hydraulic oil leaks _____

Lock the handle and test jack operation up and down. _____

Examine the jack stands and note the condition of the stand, the lifting arm, teeth, and release lever. _____
_____

_____

_____

Locate the load rating capacity for the floor jack and jack stands. _____

Describe how to properly use the floor jack and jack stands to safely raise and support a vehicle. _____
_____

_____

_____

Lab Activity 2-2

Name _____ Date _____ Instructor _____

Hoist manufacturer _____ Lifting capacity _____

Type of hoist: Above ground   In-ground   Symmetrical swing arm

　　　　　　　Asymmetrical swing arm   Drive-on

Vehicle hoist inspection

1. Examine the swing arms. Move the arms on their pivots, slide the extensions in and out, and inspect and raise the lifting pads. Note your findings.

   _____

   _____

2. Locate the instructions for using the hoist and summarize how to operate.

   _____

   _____

3. Locate any warning decals and summarize the safety precautions for using the hoist.

   _____

   _____

4. What information is necessary to use the hoist correctly and safely?

   _____

   _____

5. Examine the hoist for any signs of hydraulic oil leaks and note your findings.

   _____

   _____

6. Describe how the lift's mechanical safety mechanism operates and how to disengage the safety to lower the lift. _____

   _____

   _____

Lab Activity 2-3

Name _____ Date _____ Instructor _____

Compressed air system

1. Locate and examine compressed air outlets and hoses. Note your findings.

   _____

   _____

2. What is the shop air pressure? _____

3. Examine the quick-disconnect fittings. Note the condition of the fitting and sleeve.

   _____

   _____

4. When the air is not in use, you should turn the hose _____.

5. Describe how to safely use a blow gun to clean dirt from a part.

   _____

   _____

Lab Activity 2-4

Name _____ Date _____ Instructor _____

First aid

1. Describe the location of the first aid kit.

   _____

   _____

2. What are the basic components of the first aid kit?

   _____

   _____

3. In the event that someone in the lab begins to have a seizure, what are the steps to help that person?

   _____

   _____

4. Locate the emergency shower and eyewash station(s). Describe how to operate each.

   _____

   _____

5. With instructor permission, activate the emergency shower and eyewash station to test their operation. Note your findings.

   _____

   _____

Lab Activity 2-5

Name _____ Date _____ Instructor _____

Battery safety

1. Explain the hazards associated with automotive batteries.

   _____

   _____

2. By which hazardous material category are batteries classified? _____

3. Locate a battery either in the lab or in a vehicle. Describe battery location.

   _____

4. Is there any evidence of acid leaks and corrosion?   Yes ____ No _____

5. Describe how battery acid and corrosion can be neutralized.

   _____

   _____

6. When disconnecting a battery from the vehicle, list the proper order in which to remove and reinstall the battery cable connections.

   _____

   _____

7. When connecting a battery charger to a battery removed from the vehicle, list the steps to properly connect the battery charger.

   _____

   _____

8. What color are the high-voltage wiring and components in a hybrid vehicle?

   _____

Lab Worksheet 2-1

Name _____ Station _____ Date _____

First Aid

ASE Education Foundation Correlation

This lab worksheet addresses the following **RST** task:

1-1. Identify general shop safety rules and procedures.

Procedure

1. Describe the location of the first aid kit. _____
   _____
   _____

2. What are the basic components of the first aid kit?_____
   _____
   _____

3. If someone in the lab begins to have a seizure, what are the steps to help that person?
   _____
   _____

4. Locate the emergency shower and eye wash. Describe how to operate each.
   _____
   _____

5. With instructor permission, activate the emergency shower and eye wash to test their operation. Note your findings.
   _____
   _____

Lab Worksheet 2-2

Name _____ Station _____ Date _____

Shop Ventilation

ASE Education Foundation Correlation

This lab worksheet addresses the following **RST** task:

1-5. Utilize proper ventilation procedures for working within the lab/shop area.

Procedure

1. Explain in detail why proper ventilation of exhaust gases is important for health and safety reasons.

   _____
   _____
   _____
   _____

2. Describe how carbon monoxide causes illness or death._____

   _____
   _____
   _____
   _____

3. Inspect the exhaust ventilation equipment for your shop. Check all exhaust hoses, pipes, adaptors, and other components for wear, damage, and operation. Note your findings.

   _____
   _____
   _____
   _____

4. Demonstrate to your instructor how to properly vent exhaust in your shop.

   Instructor's check _____

Lab Worksheet 2-3

Name _____ Station _____ Date _____

Identify Marked Safety Areas

ASE Education Foundation Correlation

This lab worksheet addresses the following **RST** tasks:

1-6. Identify marked safety areas.

1-7. Identify the location and the types of fire extinguishers and other fire safety equipment; demonstrate knowledge of the procedures for using fire extinguishers and other fire safety equipment.

1-8. Identify the location and use of eye wash stations.

1-9. Identify the location of the posted evacuation routes.

Procedure

In the space below, draw a layout of your lab. Identify the locations of the following:

Marked safety areas	_____	Emergency shut-off switches	_____
Fire extinguisher locations and types	_____	Emergency shower and eye wash stations	_____
Fire blanket(s)	_____	Hazardous waste storage	_____
Fire alarms pulls	_____	SDS/MSDS sheets	_____
First aid kits	_____	Tornado safety zone	_____
Fire exits/evacuation routes	_____	Other	_____

Instructor's check _____

Lab Worksheet 2-4

Name _____ Station _____ Date _____

Fire Extinguishers

ASE Education Foundation Correlation

This lab worksheet addresses the following **RST** task:

1-7. Identify the location and the types of fire extinguishers and other fire safety equipment; demonstrate knowledge of the procedures for using fire extinguishers and other fire safety equipment.

Procedure

1. Describe the location of the fire extinguishers in the shop and note the last inspection dates.

Type of Extinguisher	Location	Inspection Date
_____	_____	_____
_____	_____	_____
_____	_____	_____
_____	_____	_____

2. Do any of the fire extinguishers need to be charged?

   No/Yes (which ones) _____

3. Watch the video located here: Public\Auto-Thompson\Video\Basics and Safety\Using a Fire Extinguisher

   a. An average fire extinguisher will discharge completely in _____ seconds.

   b. Define P.A.S.S. _____

   c. Describe how to properly use an extinguisher. _____

4. What is the best type of extinguisher for the following types of fires?

   - Cleaning solvent _____
   - Fuse box wiring _____
   - Shop rags _____
   - Gasoline _____
   - Car battery _____
   - Paper trash _____
   - Clothing _____

Lab Worksheet 2-5

Name _____ Station _____ Date _____

Fire Extinguishers Part 2

ASE Education Foundation Correlation

This lab worksheet addresses the following **RST** task:

1-7. Identify the location and the types of fire extinguishers and other fire safety equipment; demonstrate knowledge of the procedures for using fire extinguishers and other fire safety equipment.

Procedure

1. List the three most common fire classes.

   _____

   _____

   _____

2. What are the contents of each of the three most common types of fire extinguishers?

   _____

   _____

   _____

3. Why are there different extinguisher types for different types of fires?

   _____

   _____

   _____

4. What could be the consequences of using the wrong type of extinguisher on a fire?

   _____

   _____

   _____

5. Explain how to use a fire extinguisher to put out a small gasoline fire.

   _____

   _____

   _____

Lab Worksheet 2-5

Name _____ Station _____ Date _____

Fire Extinguishers Part 2

ASE Education Foundation Correlation

This lab worksheet addresses the following RST task.

1-7. Identify the location and the types of fire extinguishers and other fire safety equipment; demonstrate knowledge of the procedures for using fire extinguishers and other fire safety equipment.

Procedure

1. List the three most common fire classes.

_____

_____

_____

2. What are the contents of each of the three most common types of fire extinguishers?

_____

_____

_____

3. Why are there different extinguisher types for different types of fires?

_____

_____

_____

4. What could be the consequences of using the wrong type of extinguisher on a fire?

_____

_____

_____

5. Explain how to use a fire extinguisher to put out a small gasoline fire.

_____

_____

Lab Worksheet 2-6

Name _____ Station _____ Date _____

Personal Safety and PPE

ASE Education Foundation Correlation

This lab worksheet addresses the following **RST** tasks:

1-10. Comply with the required use of safety glasses, ear protection, gloves, and shoes during lab/shop activities.

1-11. Identify and wear appropriate clothing for lab/shop activities.

1-12. Secure hair and jewelry for lab/shop activities.

Procedure

1. Explain the importance of using proper personal protective equipment (PPE) in the automotive shop.

   _____

   _____

   _____

2. What are four examples of shop activities that pose significant danger to your eyes if safety glasses or other protective eyewear is not used?

   _____

   _____

   _____

3. Describe two examples of shop activities that require the use of hearing protection.

   _____

   _____

   _____

4. Describe five examples of shop activities in which either chemical or mechanics gloves should be worn.

   _____

   _____

   _____

5. List at least three reasons why special footwear is required in the automotive shop.

   _____

   _____

   _____

6. Describe how wearing the proper clothing is part of automotive shop safety.

   _____

   _____

   _____

7. Explain how to properly dress for working in the automotive shop.

   _____

   _____

   _____

8. Describe how leaving long hair unsecured poses a safety risk in the automotive shop.

   _____

   _____

   _____

9. Explain how to properly secure long hair for working in the automotive shop.

   _____

   _____

   _____

Instructor's check _____

Lab Worksheet 2-7

Name _____ Station _____ Date _____

SDS/MSDS

ASE Education Foundation Correlation

This lab worksheet addresses the following **RST** task:

1-15. Locate and demonstrate knowledge of material safety data sheets (MSDS).

Procedure

1. Describe the location of the SDS/MSDS book. _____

2. Why is knowing the location of the MSDS important?

   _____

   _____

3. In addition to the lab MSDS book, where else can MSDS information be found?

   _____

   _____

4. Select three different MSDS for three items used in the lab and find the following information:

   a. Product name _____   Product type   _____

   b. Hazard classification: flammable   toxic   reactive   corrosive

      If flammable, what is the flashpoint _____

      If toxic, by what sort of exposure _____

      If reactive, with what other agent(s) _____

      If corrosive, what is the pH _____

   c. Recommended PPE _____

   d. Handling and disposal requirements _____

   e. Proper first aid for exposure _____

a. Product name _____   Product type _____

b. Hazard classification: flammable   toxic   reactive   corrosive

If flammable, what is the flashpoint _____

If toxic, by what sort of exposure _____

If reactive, with what other agent(s) _____

If corrosive, what is the pH _____

c. Recommended PPE _____

d. Handling and disposal requirements _____

e. Proper first aid for exposure _____

a. Product name _____   Product type _____

b. Hazard classification: flammable   toxic   reactive   corrosive

If flammable, what is the flashpoint _____

If toxic, by what sort of exposure _____

If reactive, with what other agent(s) _____

If corrosive, what is the pH _____

c. Recommended PPE _____

d. Handling and disposal requirements _____

e. Proper first aid for exposure _____

Shop Orientation

Review Questions

1. One of the main goals of the program instructor is to teach you how to perform repairs _____ and by the manufacturer's recommended service procedures.

2. List four common areas or items that are part of the cleanup routine for your shop.

   a. _____

   b. _____

   c. _____

   d. _____

3. The primary purposes of a teaching environment are _____ and _____.

4. The primary goals of a repair shop are to repair vehicles quickly and _____ and to make a _____.

5. A new employee often must first complete a _____ period, during which his or her attendance, dependability, and initiative are closely monitored.

6. Explain why as a new employee you are not likely to be the highest-paid technician in the shop?

   _____

   _____

   _____

7. List three reasons to maintain a clean shop.

   a. _____

   b. _____

   c. _____

8. Shop cleanup often starts with making sure your _____ and the shop tools are clean, organized, and _____ to the proper location.

9. Which of the following may be part of an entry-level employee's job responsibilities?

   a. Shuttling customers

   b. Obtaining parts

   c. Shop housekeeping

   d. All of the above

10. Describe what is included in performing a PDI.

    _____

    _____

11. The idea that you will need to continue your education and training as a technician is called _____

    _____.

12. A tool that is broken or damaged can result in lost _____ for the technician and could even cause personal _____ or damage to a vehicle.

13. Explain the uses of the box end and the open end of a combination wrench.

    _____

    _____

14. When loosening or tightening a fastener, you should pull toward yourself/push away from yourself (circle the correct choice).

15. When using a wrench to loosen or tighten a fastener, use the _____ end of the wrench.

16. When trying to remove a tight bolt, a technician should use which of the following?

    a. Open-end wrench

    b. Socket and ratchet

    c. Pliers

    d. Any of the above

17. Which is not a common type of socket?

    a. 12 point

    b. 8 point

    c. 6 point

    d. Shallow

18. All of the following are common ratchet and socket drive sizes except:
    a. ¼ inch
    b. 2 inches
    c. ¾ inch
    d. ½ inch

19. Describe the care and maintenance of a ratchet to keep it in good working condition.

    _____

    _____

20. When discussing screwdriver types: *Technician A* says Torx and Phillips are similar enough that each can be used in place of the other. *Technician B* says Phillips screw heads are commonly used as plastic trim fasteners. Who is correct?
    a. Technician A
    b. Technician B
    c. Both a and b
    d. Neither a nor b

21. Never use _____ in place of the correct tool since they can damage fasteners and other parts.

22. When installing a hubcap, which type of hammer may be used?
    a. Dead blow
    b. Ball peen
    c. Rubber
    d. Brass

23. When trying to align two holes to install a fastener, a _____ may be used.

24. Click, dial, beam, and digital are all types of _____.

25. Explain how to properly set a click-type torque wrench.

    _____

    _____

26. Air-operated _____ are often used to remove lug nuts and other tight fasteners.

27. Identify the tools shown in Figure 3-1.

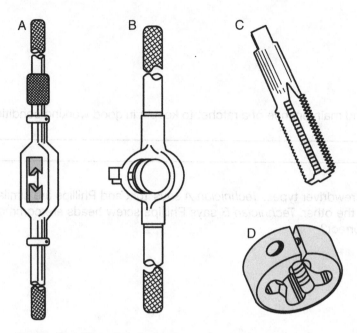

Figure 3-1 Tap and die tools.

A. _____

B. _____

C. _____

D. _____

28. When a threaded hole is so damaged that the threads cannot be repaired, a _____
_____ can often be used to fix the damage.

29. Air tools should be _____ daily and checked for proper operation each time they are used.

30. List four examples of tools commonly supplied by the shop.

    a. _____

    b. _____

    c. _____

    d. _____

31. Bench _____ are used to reshape metal and dress tools.

32. A _____ can be either petroleum- or water-based and is used to remove dirt,
oil, grease, and other substances from parts.

33. List six steps to properly connect and disconnect a battery charger.

a. _____

b. _____

c. _____

d. _____

e. _____

f. _____

34. Shop air pressure should be regulated to:

a. 60 psi

b. 90 psi

c. 120 psi

d. It is not regulated

35. _____ and _____ covers are used to help protect the vehicle during service.

36. Identify the parts of the bolt shown in Figure 3-2.

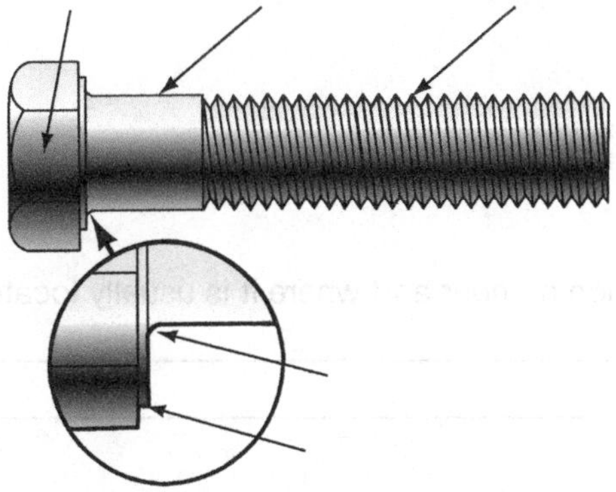

Figure 3-2 Bolt part ID.

37. Describe how to properly measure a bolt to determine its dimensions.

_____

_____

38. List seven steps to properly maintain a torque wrench.

a. _____

b. _____

c. _____

d. _____

e. _____

f. _____

g. _____

39. Explain why is it important to always lower the lift onto the safety mechanism before working under the vehicle.

   _____

   _____

40. To use a floor jack to raise a vehicle, you must turn the handle _____.

41. List the eight steps to lift and support a vehicle with a floor jack and jack stands.

   a. _____

   b. _____

   c. _____

   d. _____

   e. _____

   f. _____

   g. _____

   h. _____

42. Define the following acronyms:

   a. FWD

   b. RWD

   c. 4WD

   d. AWD

43. Describe the vehicle identification number and where it is usually located on a vehicle.

   _____

   _____

44. What type of information is contained on the vehicle emission control information (VECI) decal?

   _____

   _____

45. Which of the following is often contained on a door decal?

   a. Tire pressure

   b. Paint codes

   c. Gross vehicle weight

   d. All of the above

46. What is the purpose of the primary and secondary hood release mechanisms?

   _____

   _____

47. Describe six components of a service order.

    a. _____

    b. _____

    c. _____

    d. _____

    e. _____

    f. _____

Activities

1. Identify the tools shown in Figure 3-3 (A through G).

A. _____

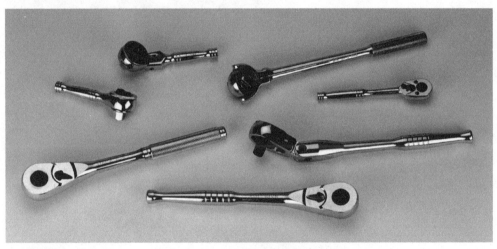

B. _____

C. _____

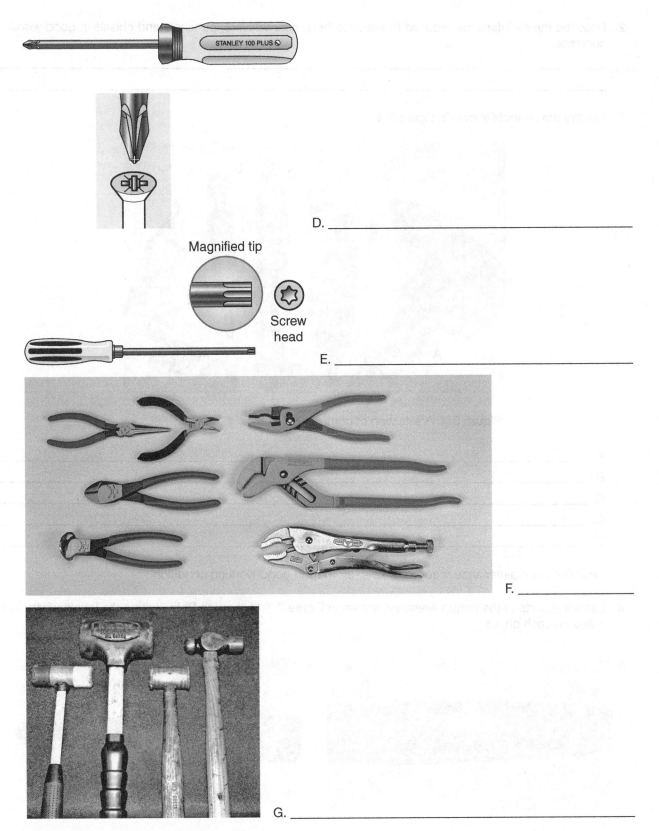

Magnified tip

Screw head

D. _____

E. _____

F. _____

G. _____

Figure 3-3 A selection of basic hand tools.

2. Describe the maintenance required to keep ratchets, screwdrivers, punches, and chisels in good working condition.

_____

_____

3. Identify the air tools shown in Figure 3-4.

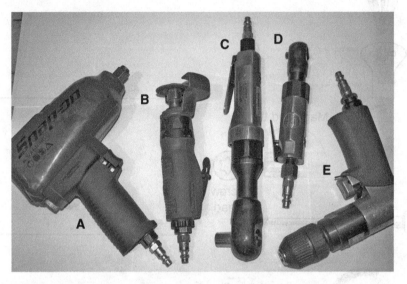

Figure 3-4 A selection of common air tools.

A. _____

B. _____

C. _____

D. _____

E. _____

Describe the maintenance required to keep air tools in good working condition.

4. Label the parts of the torque wrenches shown in Figure 3-5 (A through D) by writing the labels where indicated on each photo.

A.

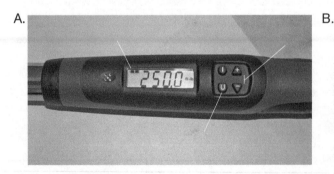

B.

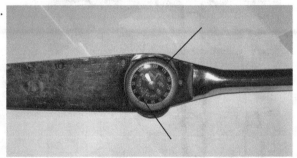

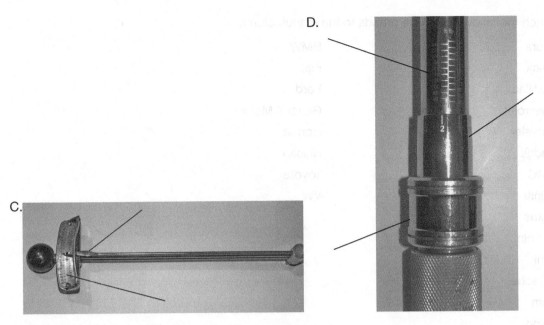

Figure 3-5 Click, beam, and digital torque wrenches.

Explain how to properly set and store a click-type torque wrench.

_____

_____

5. Identify the lifts used in your lab. Describe each lift and how it operates. If there are several lifts of the same manufacturer and type, just describe one of the lifts.

Lift 1. _____

_____

Lift 2. _____

_____

Lift 3. _____

_____

6. Match the following vehicle brands to the manufacturer.

Acura	BMW
Buick	Fiat
Cadillac	Ford
Chevrolet	General Motors
Chrysler	Honda
Dodge	Nissan
GMC	Toyota
Infiniti	VW
Lexus	
Lincoln	
Mini	
Porsche	
Ram	
Scion	

Lab Activity 3-1

Name _____ Date _____ Instructor _____

Hand Tools

1. Using your toolkit, locate the English and metric sizes that are close to matching each other in size.

   ½ inch = _____ mm   14 mm = _____ inch   ⅝ inch = _____ mm

   10 mm = _____ inch   ¾ inch = _____ mm   8 mm = _____ inch

   a. Why is it important to use the correct size socket or wrench on a fastener?

   _____

   _____

   b. Describe how to determine if a fastener is English or metric.

   _____

   _____

2. Using a selection of fasteners supplied by your instructor, match the correct size socket to the head of the bolt.

   Bolt 1 _____   Bolt 2 _____   Bolt 3 _____

   Bolt 4 _____   Bolt 5 _____   Bolt 6 _____

Lab Activity 3-2

Name _____ Date _____ Instructor _____

Bench grinder use

1. Label the parts of the bench grinder shown in Figure 3-6 (A through D).

Figure 3-6 Bench grinder.

A. _____

B. _____

C. _____

D. _____

2. Examine the eye shields, grinder wheels, and power cord before use. Note your findings.

_____

_____

3. Turn the grinder on and listen for any noise that may indicate a problem with the grinder. Examples can be growling from the bearings or from loose debris in the wheel covers. Do not use the grinder if there is any damage to the grinder wheels. Note your findings. _____

_____

4. With the grinder off, make sure the work rest is tight and positioned close to the wheel. Turn the grinder on and place a worn chisel on the work rest. The chisel should be placed as shown in Figure 3-7.

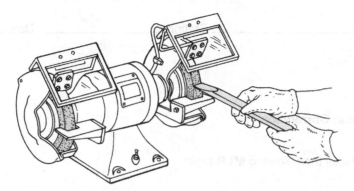

Figure 3-7 Positioning of chisel on bench grinder for sharpening.

5. Move the chisel back and forth across the grinder wheel until properly dressed.

Lab Worksheet 3-1

Name _____ Station _____ Date _____

External Thread Repair

ASE Education Foundation Correlation

This lab worksheet addresses the following **MLR** task:

1.A.6. Perform common fastener and thread repair, to include: remove broken bolt, restore internal and external threads, and repair internal threads with thread insert. **(P-1)**

Procedure

1. To repair external threads, such as on a bolt or a stud, first determine the fastener diameter and thread pitch.

   Measure the diameter using a machinist's ruler: diameter _____

   Measure the thread pitch using a thread pitch gauge: thread pitch _____

2. Select the correct die. Place a small amount of cutting oil on the fastener and carefully begin to thread the die on the threads. Turn the die approximately one-quarter of a turn and then back off slightly to allow any shavings to dislodge from the die. Continue to thread the die along the fastener until the threads are restored.

   Instructor's check _____

3. Remove the die and clean the die and the bolt or stud. Remove all shavings and cutting oil.

   Instructor's check _____

Lab Worksheet 3-2

Name _____ Station _____ Date _____

Internal Thread Repair

ASE Education Foundation Correlation

This lab worksheet addresses the following **MLR** task:

1.A.6. Perform common fastener and thread repair, to include: remove broken bolt, restore internal and external threads, and repair internal threads with thread insert. **(P-1)**

Procedure

1. To repair internal threads, such as a bolt hole, first determine the correct fastener size that fits the hole. Measure the thread diameter and thread pitch of a bolt used in a similar hole.

   Measure the diameter using a machinist's ruler: diameter _____

   Measure the thread pitch using a thread pitch gauge: thread pitch _____

2. Select the correct size tap. Place a small amount of cutting oil on the tap and carefully begin to thread the tap into the hole. It is vital that the tap remain properly positioned to the hole to prevent damage to the threads and to the tap. Turn the tap approximately one-quarter of a turn and then back off slightly to allow any shavings to dislodge from the tap. Continue to thread the tap into the hole until the threads are restored.

   CAUTION: If the tap seats in the bottom of a blind hole, do not tighten the tap into the hole.

   Overtightening the tap could cause it to break off in the hole. Verify you are using the correct type of tap for the hole.

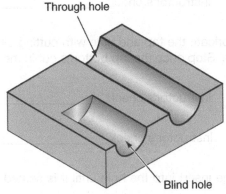

Through hole

Blind hole

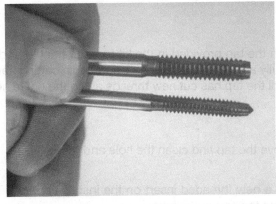

3. Remove the tap and clean the tap and the bolt hole. Remove all shavings and cutting oil.

   Instructor's check _____

Lab Worksheet 3-3

Name _____ Station _____ Date _____

Threaded Insert

ASE Education Foundation Correlation

This lab worksheet addresses the following **MLR** task:

1.A.6.   Perform common fastener and thread repair, to include: remove broken bolt, restore internal and external threads, and repair internal threads with thread insert. **(P-1)**

Procedure

1. To repair internal threads that have been severely damaged, a threaded insert may be required. To repair internal threads, such as a bolt hole, first determine the correct fastener size that fits the hole. Measure the thread diameter and thread pitch of a bolt used in a similar hole.

   Measure the diameter using a machinist's ruler: diameter _____

   Measure the thread pitch using a thread pitch gauge: thread pitch _____

2. Obtain an internal thread repair kit, often called a Heli-Coil, from your instructor. The kit often contains an oversized drill bit, an oversize tap, threaded inserts, and an installation tool. For this repair, what size drill bit is required? Drill bit size _____

3. Using the supplied drill bit, carefully drill the hole oversize to accommodate the insert. Use a blow gun to remove all debris from the hole.

   Instructor's check _____

4. Attach the tap provided in the kit to a tap handle or driver. Lubricate the tap and hole with cutting oil and carefully align the tap to the hole and begin cutting new threads. Stop once the tap bottoms out in the hole or until the tap has cut new threads along the entire hole.

   Instructor's check _____

5. Remove the tap and clean the hole and tap.   Instructor's check _____

6. Place a new threaded insert on the installation tool. Thread the insert into the hole until it is seated or it reaches the end of the hole.

   Instructor's check _____

7. Remove the locking tang if applicable.   Instructor's check _____

8. Remove the insert tool and test the new threads using an appropriate fastener.   Instructor's check _____

Lab Worksheet 3-4

Name _____ Station _____ Date _____

Remove a Broken Bolt

ASE Education Foundation Correlation

This lab worksheet addresses the following **MLR** task:

1.A.6. Perform common fastener and thread repair, to include: remove broken bolt, restore internal and external threads, and repair internal threads with thread insert. **(P-1)**

Procedure

1. To remove a bolt broken off in a bolt hole, first attempt to thread the bolt out using a pick or similar tool. If the bolt is stuck in the hole, it will need to be drilled out and an extractor used. If possible, flatten out the top of the broken bolt with a file or grinder. Use a centering punch to punch a small dent in center of the top of the bolt. This will be used to position the drill bit.

   Instructor's check _____

2. Select a small drill bit, such as a ⅛-inch bit as a pilot hole bit. Apply a drop of cutting oil to the bit. Keep the bit centered and aligned over the bolt and begin to drill into the bolt.

   Instructor's check _____

3. Once the pilot hole is through the bolt, use a larger bit to cut out the majority of the inside of the bolt. Use a bit one or two sizes smaller than the actual diameter of the bolt.

   Drill bit size_____

4. Once the bolt has been drilled out, use a left handed drill bit or screw extractor to remove the bolt.

   Instructor's check _____

Screw extractor

Broken bolt with hole drilled in the middle

5. Once the bolt has been removed, inspect the internal threads in the hole and repair as necessary.

   Instructor's check _____

6. Once finished, put away all tools and equipment and clean up the work area.

   Instructor's check _____

Lab Worksheet 3-5

Name _____ Station _____ Date _____

Compressed Air System

ASE Education Foundation Correlation

This lab worksheet addresses the following **RST** task:

1-2. Utilize safe procedures for handling of tools and equipment.

Procedure

1. Examine the shop's compressed air outlets and hoses. Note your findings.
   _____
   _____

2. What is the shop air pressure? _____

3. Examine the quick-disconnect fittings. Note the condition of the fitting and sleeve.
   _____
   _____

4. When the air is not in use, the air line/hose should be left: _____ On _____ Off

5. Describe how to safely use a blow gun to clean dirt from a part. _____
   _____
   _____

Lab Worksheet 3-6

Name _____ Station _____ Date _____

Chemicals

ASE Education Foundation Correlation

This lab worksheet addresses the following **RST** task:

1-2. Utilize safe procedures for handling of tools and equipment.

Procedure

1. Compile a list of commonly used chemicals in the auto lab. Include at least five chemicals.

   _____

   _____

   _____

2. Which of these chemicals have special handling and disposal requirements?

   _____

   _____

3. How and where is waste coolant stored for your lab?

   _____

   _____

4. How and where is waste oil stored for your lab?

   _____

   _____

5. Why are waste coolant and waste oil products stored separately?

   _____

   _____

6. Describe the type of solvent cleaning system used in your lab.

   _____

   _____

7. How is waste solvent handled and disposed of?

   _____

   _____

Lab Worksheet 3-7

Name _____ Station _____ Date _____

Jacking and Supporting a Vehicle

ASE Education Foundation Correlation

This lab worksheet addresses the following **RST** task:

1-3. Identify and use proper placement of floor jacks and jack stands.

Description of Vehicle

Name _____ Date _____

Year _____ Make _____ Model _____

Procedure

1. Locate and note the recommended lifting and jacking points for this vehicle.
   _____
   _____

2. Before lifting, the transmission should be placed in _____

3. Place a set of wheel chocks against the front and back of a tire on the axle that is not being raised.

4. Position the floor jack under a lift point and raise the vehicle slightly. Have your instructor check before proceeding.   Instructor's check _____

5. Continue to raise the vehicle until the wheel(s) are off the floor. Position the jack stand(s) under the vehicle where it is to be supported. If using two jack stands, ensure that both are set to the same height. Note the location for the jack stands.

6. Slowly release the floor jack so the vehicle settles onto the jack stand.
                                                   Instructor's check _____

7. Recheck that the vehicle is safely contacting the jack stand and that the stands are fully supporting the vehicle.
                                                   Instructor's check _____

8. To lower the vehicle, position the floor jack under a lift point. Raise the jack until the vehicle lifts up off of the jack stands. Remove the jack stands and slowly lower the vehicle to the ground.

Instructor's check _____

9. Place the transmission back into _____. Remove the wheel chocks.

10. Explain why it is important to locate the correct lifting points when using a floor jack. _____
   _____
   _____

11. Why is it important for the jack stands to be set to the same height?_____
   _____
   _____

12. Explain why you should never work on a vehicle supported only with a floor jack. _____
   _____
   _____
   _____

Instructor's check _____

Lab Worksheet 3-8

Name _____ Station _____ Date _____

Vehicle Lift Inspection

ASE Education Foundation Correlation

This lab worksheet addresses the following **RST** task:

1-4. Identify and use proper procedures for safe lift operation.

Description

Hoist manufacture _____   Lifting capacity _____

Type of hoist: Above ground   two post   In-ground two post   Symmetrical swing-arm

Asymmetrical swing-arm   Drive-on   In-ground single post

Procedure

1. Examine the swing arms. Move the arms on their pivots, slide the extensions in and out, and inspect and raise the lifting pads. Note your findings.
    _____
    _____

2. Locate the instructions for using the hoist and summarize its operation. _____
    _____
    _____

3. Locate any warning decals and summarize the safety precautions for using the hoist._____
    _____
    _____

4. What additional information is necessary to use the hoist correctly and safely?_____
    _____
    _____

5. Examine the hoist for any signs of hydraulic oil leaks. Note your findings._____
    _____
    _____

6. Describe how the lifts mechanical safety mechanism operates and how to disengage the safety to lower the lift.
    _____
    _____

Lab Worksheet 3-9

Name _____ Station _____ Date _____

Fasteners

ASE Education Foundation Correlation

This lab worksheet addresses the following **RST** tasks:

2-1. Identify tools and their usage in automotive applications.

2-2. Identify standard and metric designation.

2-3. Demonstrate safe handling and use of appropriate tools.

Procedure

Using a selection of bolts supplied by your instructor, determine the correct thread pitch, length, and diameter for each.

1. TP _____ Diameter _____ Length _____

2. TP _____ Diameter _____ Length _____

3. TP _____ Diameter _____ Length _____

4. TP _____ Diameter _____ Length _____

5. TP _____ Diameter _____ Length _____

6. TP _____ Diameter _____ Length _____

7. What will happen if the incorrect bolt thread is inserted and tightened into a threaded hole?

_____

_____

Lab Worksheet 3-10

Name _____ Station _____ Date _____

Bench Grinder Use

ASE Education Foundation Correlation

This lab worksheet addresses the following **RST** task:

2-3. Demonstrate safe handling and use of appropriate tools.

Procedure

1. Ensure that all guards and shields are in place before attempting to use a bench grinder (see below).

BENCH GRINDER

Eye shield

Fine wheel

Work rest

Medium wheel

2. Examine the eye shields, grinder wheels, and power cord before use. Note your findings.
_____
_____

3. Turn the grinder on and listen for any noise that may indicate a problem with the grinder. Examples can be growling from the bearings or of debris lose in the wheel covers. Do not use the grinder if there is any damage to the grinder wheels. Note your findings.
_____
_____

4. With the grinder off, make sure the work rest is tight and positioned close to the wheel. Turn the grinder on and place a worn chisel on the work rest. The chisel should be placed as shown in the image above.

Instructor's check _____

5. Move the chisel back and forth across the grinder wheel until properly dressed.

Instructor's check _____

Lab Worksheet 3-11

Name _____ Station _____ Date _____

Hand Tools

ASE Education Foundation Correlation

This lab worksheet addresses the following **RST** task:

2-3. Demonstrate safe handling and use of appropriate tools.

Procedure

1. Using your tool kit, locate the English and metric sizes that are close to matching each other in size.

   ½ inch = _____ mm   14 mm = _____ inch   ⅝ inch = _____ mm

   10 mm = _____ inch   ¾ inch = _____ mm   8 mm = _____ inch

2. Why is it important to use the correct size socket or wrench on a fastener? _____
   _____
   _____

3. Describe how to determine if a fastener is English or metric. _____
   _____
   _____

4. Using a selection of fasteners supplied by your instructor, match the correct size socket to the head of the bolt.

   Bolt 1 _____   Bolt 2 _____   Bolt 3 _____

   Bolt 4 _____   Bolt 5 _____   Bolt 6 _____

   Nut 1 _____   Nut 2 _____   Nut 3 _____

Lab Worksheet 3-12

Name _____ Station _____ Date _____

Tools and Equipment

ASE Education Foundation Correlation

This lab worksheet addresses the following **RST** task:

2-4. Demonstrate proper cleaning, storage, and maintenance of tools and equipment.

Procedure

1. Explain why proper tool cleaning, maintenance, and storage is important for both the technician and the shop.
   _____
   _____
   _____
   _____
   _____

2. Describe how to properly clean basic hand tools. _____
   _____
   _____

3. Which basic hand tools may require periodic maintenance to be kept in good working condition?
   _____
   _____

4. Describe the tool storage methods for you class. _____
   _____
   _____
   _____

5. Describe the tool maintenance program for your class. _____
   _____
   _____
   _____

Instructor's check _____

Lab Worksheet 3-13

Name _____ Station _____ Date _____

Precision Measuring

ASE Education Foundation Correlation

This lab worksheet addresses the following **RST** task:

2-5. Demonstrate proper use of precision measuring tools (i.e., micrometer, dial-indicator, dial-caliper).

Procedure

Using a selection of parts supplied by your instructor, measure each and record your readings.

Micrometer: measure each part, note if using an English or metric micrometer, and record your readings.

1. Item _____ English/Metric Measurement _____
2. Item _____ English/Metric Measurement _____
3. Item _____ English/Metric Measurement _____
4. Item _____ English/Metric Measurement _____

Dial indicator: measure the runout or play and record your readings.

5. Item _____ English/Metric Measurement _____
6. Item _____ English/Metric Measurement _____

Dial caliper: measure the runout or play and record your readings.

7. Item _____ English/Metric Measurement _____
8. Item _____ English/Metric Measurement _____

Instructor's check _____

Lab Worksheet 3-14

Name _____ Station _____ Date _____

Vehicle Information

ASE Education Foundation Correlation

This lab worksheet addresses the following **RST** task:

3-1. Identify information needed and the service requested on a repair order.

Description of Vehicle

Name _____ Date _____

Year _____ Make _____ Model _____

Procedure

1. Locate and record the VIN: VIN _____

    Location _____

2. Using the decal(s) in the door jamb, locate and record the following information:

    Build date _____   Gross weight _____

    Tire pressure _____   Tire size _____

3. Locate the vehicle emission control identification (VECI) decal and record the following information:

    Emission year _____   Engine size _____

    Tier/Bin _____

    Installed emission devices _____

4. For General Motors vehicles, locate the SPO tag and note its location:

    SPO tag location _____

    What is the purpose of the SPO tag? _____

Lab Worksheet 3-15

Name _____ Station _____ Date _____

VIN ID

ASE Education Foundation Correlation

This lab worksheet addresses the following **RST** task:

3-1. Identify information needed and the service requested on a repair order.

Description of Vehicle

Name _____ Date _____

Year _____ Make _____ Model _____

Procedure

1. Locate and record the vehicle identification number.

   VIN _____

   Location _____

2. What type of information is contained in the VIN? _____
   _____
   _____

3. Using service information, determine the following based on the VIN:

   a. Model year_____

   b. Country of manufacture _____

   c. Engine size _____

   d. Manufacturer/Division _____

   e. Body/Platform type _____

   f. Serial number _____

   g. Check digit _____

4. Examine the vehicle and note other locations where the VIN can be found:

   a. _____

   b. _____

   c. _____

   d. _____

5. Why is the VIN located in more than one place? _____

   _____

   _____

Lab Worksheet 3-16

Name _____ Station _____ Date _____

Fender Covers

ASE Education Foundation Correlation

This lab worksheet addresses the following **RST** task:

3-2. Identify purpose and demonstrate proper use of fender covers, mats.

Procedure

1. Explain the purpose of using fender covers and other covers when working on a vehicle.

   _____

   _____

   _____

   _____

2. Explain why it is important for the technician and the shop to protect the vehicle from damage during service.

   _____

   _____

   _____

   _____

3. List the areas of the vehicle that should be covered and protected when brought in for service.

   _____

   _____

   _____

4. Demonstrate to your instructor how to properly protect a vehicle using covers supplied for the class.

   Instructor's check _____

Lab Worksheet 3-17

Name _____ Station _____ Date _____

Concern, Cause, and Correction

ASE Education Foundation Correlation

This lab worksheet addresses the following **RST** task:

3-3. Demonstrate use of the three Cs (concern, cause, and correction).

Procedure

1. What is the concern on this sample repair order?
   _____
   _____

2. What is the cause of the concern on this sample repair order?
   _____
   _____

3. What is the correction on this sample repair order?
   _____
   _____

Lab Worksheet 3-18

Name _____ Station _____ Date _____

Complete Work Order

ASE Education Foundation Correlation

This lab worksheet addresses the following **RST** tasks:

3-3. Demonstrate use of the three Cs (concern, cause, and correction).

3-4. Review vehicle service history.

3-5. Complete work order to include customer information, vehicle identifying information, customer concern, related service history, cause, and correction.

Description of Vehicle

Year _____ Make _____ Model _____

Engine _____ AT MT CVT PSD (circle which applies)

Procedure

1. Begin a work order for a customer, include the following information:

   Customer name and contact information _____

   Vehicle information _____

   Instructor's check _____

2. Determine the customer's concern or service needed. _____
   _____
   _____

3. Identify and note any of the following:

   Applicable TSBs _____
   _____

   Recalls _____
   _____

   Service history _____
   _____
   _____

4. Prepare an estimate for repairs and attach.

Instructor's check _____

5. Once the vehicle has been inspected, note the cause(s) for the customer's concern. _____

_____

_____

6. Based on the concern and the cause, what is/are the recommended actions to correct the concern?

_____

_____

_____

Lab Worksheet 3-19

Name _____ Station _____ Date _____

Using Service Information

ASE Education Foundation Correlation

This lab worksheet addresses the following **RST** task:

3-5. Complete work order to include customer information, vehicle identifying information, customer concern, related service history, cause, and correction.

Description of Vehicle

Year _____ Make _____ Model _____

VIN: _____ Engine _____

Procedure

1. Wheel lug torque specification _____

2. Cylinder firing order _____

3. Drive belt routing (draw picture)

4. Engine oil capacity _____

5. Recommended coolant type _____

6. Labor time to replace front brake pads _____

7. Labor time to replace the fuel pump _____

8. Locate a technical service bulletin (TSB) of your choice; record the TSB number, TSB date, and a brief description of the reason for the TSB.

   _____

   _____

   _____

CHAPTER 4

Basic Technician Skills

Activities

I. Employability Skills

1. Describe what professionalism means to you.

   _____

   _____

2. Define your idea of work ethic.

   _____

   _____

3. How would other people describe your work ethic?

   _____

   _____

4. Motivation and initiative are important for auto technicians; explain what motivation and initiative mean to you.

   _____

   _____

5. Define the following terms as an employer's expectations of an employee.

Positive attitude

_____

Sense of responsibility

_____

Initiative

_____

Flexibility

_____

Dependability

_____

Productivity

_____

Enthusiasm

_____

6. Develop a list of several personal intrinsic and extrinsic motivational factors.

_____

_____

_____

II. Oral Communication

1. Why is it important to have good oral communication skills in the workplace?

_____

_____

2. What can result from not being able to communicate clearly and professionally when working with customers?

_____

_____

3. Define _slang_.

_____

_____

4. List several slang terms you use in everyday conversation and the corresponding standard English meanings for the terms.

_____

_____

_____

III. Nonverbal Communication

1. Describe three types of nonverbal forms of communication.

   _____

   _____

2. When dealing with others in a professional environment, what forms of nonverbal communication should you be careful to avoid?

   _____

   _____

IV. Reading and Writing Skills

1. Technicians must be able to read, understand, and apply what they read to perform tests and repairs. Following are words commonly used in technical service information with which you need to be familiar. Using a separate piece of paper, define each of the following verbs:

   Adjust _____   Align _____

   Analyze _____   Assemble _____

   Balance _____   Bleed _____

   Charge _____   Check _____

   Clean _____   Correct _____

   Determine _____   Diagnose _____

   Disassemble _____   Discharge _____

   Evacuate _____   Flush _____

   Hone _____   Inspect _____

   Locate _____   Measure _____

   Perform _____   Purge _____

   Remove _____   Replace _____

   Resurface _____   Service _____

   Test _____   Torque _____

   Verify _____

2. The following paragraph is based on actual service information; read the paragraph and summarize the basic concept.

   "When the ignition is switched ON, the transponder embedded in the key is energized by the exciter coils surrounding the ignition lock cylinder. The energized transponder transmits a unique signal value, which is received by the theft deterrent control module (TDCM). The TDCM compares this signal value to a value stored in read-only memory (ROM) as the learned key code. The TDCM sends a randomly generated number to the transponder, which is called a challenge. Both the transponder and the TDCM perform a calculation on the challenge. If the calculations match, the TDCM sends the fuel enable password via the serial data circuit to the engine control module (ECM). The signal is sent over the controller area network (CAN) high-speed data bus. If either the transponder's unique value or the calculation to the challenge is incorrect, the TDCM will send the fuel disable password to the ECM via the serial data circuit."

List any words used with which you are unfamiliar:

_____

Using a dictionary, locate the words you listed above as unfamiliar and define each in your own words.

_____

_____

_____

Summarize the paragraph in your own words.

_____

_____

_____

_____

_____

V. Penmanship

Examine the repair order example in Figure 4-1 to answer the following:

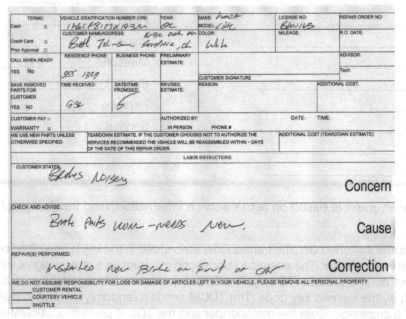

Figure 4-1 Poorly written repair order.

1. What is the customer's name? _____

2. Customer address? _____

3. Vehicle year, make, and model? _____

4. Customer concern? _____

5. Action required to determine cause of concern? _____

   _____

6. If you were the customer presented with the RO, what would be your reaction?

   _____

   _____

7. Imagine that you are a judge or a member of a jury and were making a decision about a disputed repair based on this RO. Describe your reaction to the RO.

   _____

   _____

   _____

   _____

VI. Spelling

Being able to properly spell, especially automotive terms, is an important skill for communicating with co-workers and customers. In the following example sentences, identify (circle) the misspelled words.

1. Customer says her breaks are making noise.

2. The custemer says the steering wheel is shaking when braking.

3. Rotated tires and torked wheels.

4. Replaced spark plugs, air filter, and fule filter.

5. Performed too wheel alingment, rotated and balanced the tires.

6. Check breaks, making noise when stoping.

7. Left front shok leeking.

VII. Grammar

Read the following passage and answer the questions for each section.

(1) Torque converter's transfer engine torque from the engines crankshaft to the transmission, (2) doing the need away for a mechanical clutch too couple the transmission too the engine. (3) Modern torque converters get this power transfer by fluid using the following major component—the cover or shell, impeller, turbine, and stator. (4) All late model torque converter's include a lock-up clutch, to improve vehicle's fuel economy. (5) Good diagnosis of the customers complaints is needed since some complaints can be masked by faults in other powertrain systems.

1. Choose the best correction for section 1 of the passage.

   a. Torque converters transfer engine torque from the engines' crankshaft to the transmission.

   b. Torque converters transfer engine torque from the engine's crankshaft to the transmission.

   c. Torque converters transfer engine torque from the engine crankshaft to the transmission.

   d. Torque converter's transfer engine torque from the engine's crankshaft to the transmission.

   e. No correction necessary.

2. Choose the best correction for section 2 of the passage.

   a. Doing away for a mechanical clutch to couple the transmission to the engine.

   b. This does away with a mechanical clutch to couple the transmission to the engine.

   c. This, does away a mechanical clutch to couple the transmission to the engine.

   d. Doing away with a mechanical clutch, to couple the transmission to the engine.

   e. No correction necessary.

3. Choose the best correction for section 3 of the passage.

   a. Modern torque converters, get this power transfer by using the following major component, the cover or shell, impeller, turbine, and stator.

   b. Modern torque converters, transfer fluid using the following major components: the cover or shell, impeller, turbine, and stator.

   c. Modern torque converters get this power transfer by using the following major components: the cover or shell, impeller, turbine, and stator.

   d. Modern torque converters get this power transfer, using the following major component—the cover or shell, impeller, turbine, and stator.

   e. No correction necessary.

4. Choose the best correction for section 4 of the passage.

   a. All late-model torque converters include a lock-up clutch to improve vehicle fuel economy.

   b. All late-model torque converter's include a lock-up clutch; to improve vehicle's fuel economy.

   c. All late model torque converter's include a lock-up clutch: to improve the vehicles fuel economy.

   d. All late model torque converters include a lock-up clutch, used to improve vehicle fuel economy.

   e. No correction necessary.

5. Choose the best correction for section 5 of the passage.
   a. Good diagnosis of the customer's complaint is needed, since some complaints, can be masked by faults in other powertrain systems.
   b. Proper diagnosis of the customer's complaints are needed, since some complaints can be masked by faults in other powertrain systems.
   c. Proper diagnosis of the customer's complaint is needed since some complaints can be masked by faults in other powertrain systems.
   d. Good diagnosis of the customer's complaint's is needed since some complaints can be masked by faults in other powertrain systems.
   e. No correction necessary.

VIII. Math Skills

A. Fractions

A fraction is a mathematical way to represent a division of a whole in parts. An example of the fraction ¼ can be shown as in Figure 4-2. The parts of a fraction are the numerator, the number on top of the line, and the denominator, the number below the line. The numerator represents the number of parts of the whole that are chosen and the denominator defines the total number of parts.

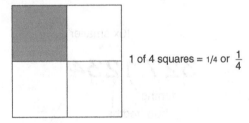

1 of 4 squares = 1/4 or $\frac{1}{4}$

Figure 4-2

1. Proper fractions like ¼ have a numerator that is smaller than the denominator. When working with SAE fasteners, you will use fractions such as ¼ inch, ⅜ inch, ½ inch, ¾ inch, and similar sizes.

Improper fractions, such as 7/3, have the numerator equal to or larger than the denominator. This type of fraction is not often used in automotive applications.

Mixed fractions, such as 2 ½, contain a whole number and a proper fraction. Measurement of bolt lengths, drum brakes, and wheelbase sizes will often use mixed fractions.

Sometimes working with fractions requires you to reduce a fraction. For example, you may measure a bolt that is 14/16 of an inch long. Since both the numerator and denominator are even numbers, both can be reduced to smaller numbers. This is done by factoring. Factoring is used to reduce fractions by listing the prime factors of both the numerator and the denominator. In our example of 14/16, the prime factors of 14 are 1, 2, and 7. The prime factors of 16 are 1, 2, 4, and 8. Since 7 goes into 14 twice and 8 goes into 16 twice, you can reduce 14/16 by writing $\frac{2 \times 7 = 14}{2 \times 8 = 16}$. Since the 2s cancel out, you are left with ⅞.

2. Reduce the following fractions to their smallest form.

   6/8, 12/16, 10/32, 2/4, 8/32 ______________________________

3. Place the following fractions in order from smallest to largest.

   13/16, 1/4, 5/8, 9/16, 7/32, 11/16, 1/2, 5/16, 9/32, 3/8, 7/16, 3/4, 7/8, 15/16 ______________________________

   Fractions can also be converted into decimals, as discussed next.

B. Decimals

A decimal is another method of expressing a portion of a whole, similar to a fraction, except unlike a fraction, a decimal does not express the part compared to a number of the whole. Whereas a fraction such as ¼ means that there are four parts in the whole, a decimal is used to show the parts in tenths, hundredths, thousandths, or more. This is based on the place value of the number. Place value determines a number's value on either side of a decimal point. A whole number, such as 327, is represented by the ones column, tens column, and one hundreds column, with 3 being in the hundreds, 2 in the tens, and 7 in the ones, as shown in Figure 4-3.

Figure 4-3

To show a portion of the whole number, decimals use numbers to the right of a decimal point, as shown in Figure 4-4. The first number to the right of the decimal point has the value of tenths, the second number is hundredths, the third is thousandths, and so on. Common measurements in automotive applications use the hundredths and thousandths place, and occasionally even the ten thousandths place.

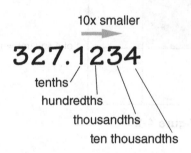

Figure 4-4

To convert a fraction into a decimal, simply divide the numerator by the denominator. For example, to convert ¼ into decimal, divide 1 by 4. If using a calculator, input 1 ÷ 4. Solve the following problems.

1. Find the sum of 21.45, 13.08, and 91.16. _____

2. Subtract 37.18 from 119.05. _____

C. Percentages

A percentage is another method to represent a portion of the whole and is also used to indicate numbers greater than the whole. Percentages are either shown as a number and percent symbol, such as 50%, or written out as 50 percent. This example, 50%, means 50 out of 100. Percentages are useful when working with money, such as figuring discounts and markup.

Solving some percentage problems, such as 10% off a sale price, is easy. If an item is on sale for $89.95, simply move the decimal point one place to the left to find the price after the discount. In this example, $89.95 is discounted $8.99, or 1/10 of the original price. Two more ways to solve this can be done with a calculator. If the calculator has a % key, you can enter 89.95 and multiply times 10% to find the discounted amount. You can also multiply 89.95 times 90% to find the final price after the discount. If the calculator does not have a % key, you can still find the percentage by multiplying 89.95 by .10 to find the discount amount or by .90 to find the total price after the discount. By using .10 or .90, you are multiplying by the value of the percentage. By using .10, this is equivalent to 1/10 or 10% of 1.

Percentages are also used to calculate markup. Markup is the difference between the sale price and what the item actually cost. For example, if you buy a drink from a vending machine for $1 but the cost of the drink to the vendor is only $0.25, then the markup on the drink is 400% since the drink sells for four times its cost.

Solve the following problems.

1. A part retails for $128.67, but the shop receives a 20% discount. What is the cost of the part to the shop?

   _____

2. A part that cost $47.90 is resold with a markup of 40%. At what price is the part sold?

   _____

3. Sales tax of 6.75% is added to all parts and labor. Find the sales tax for the following repair amounts: $19.95, $79.88, $212.45, $420.16. _____

D. Ratios

A ratio provides a mathematical relationship between two items and is written as item A in relation to item B, or A:B. Common examples include gear ratios, rocker arm ratios, braking ratios, and such proportions as mixing water and coolant in a 50:50 ratio. Gear ratios are written as 2:1 and 3.5:1, where the first number represents the number of turns of the driving gear and the second number is the turns of the driven gear, as shown in Figure 4-5.

Gear ratio = 2 to 1
(2 revolutions input to
1 revolution output)

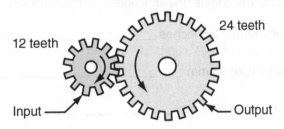

24 teeth

12 teeth

Input — Output

Figure 4-5

Solve the following problems.

Determine the gear ratio based on the following driving to driven gear.

1. Driving gear = 10 teeth, driven gear = 15 teeth _____

2. Driving gear = 20 teeth, driven gear = 50 teeth _____

3. Driving gear = 18 teeth, driven gear = 12 teeth _____

E. Area and Volume

Area is a two-dimensional measurement of an object's length and width. Surface area is calculated by multiplying the length and width of an object. While it is not often, you will need to find the area in automotive applications, it is a measurement with which you should be familiar. One example of using area is finding the contact patch size of the tire on the ground. Obtain a contact patch imprint by placing a piece of paper under a tire and allowing the vehicle weight to settle on the tire. Next, remove the paper and measure the size of the imprint of the tire. Calculate the contact patch area and multiply your answer by the recommended tire inflation pressure. Compare your answer to the maximum tire load information located on the tire sidewall.

1. What is the size of the tire's contact patch? _____

2. What is the tire's recommended inflation pressure? _____

3. What is the maximum load rating of the tire? _____

4. How do tire size and pressure affect its load-carrying capacity? _____

Volume is the three-dimensional measurement of length, width, and height. Volume measurement is used when discussing engine size, specifically the engine displacement. This number refers to the volume of air that all of the cylinders can hold. To determine the volume of a cylinder use: cylinder volume = pi/4 × bore^2 × stroke. A slightly simpler method is to use 0.785 × bore^2 × stroke because 0.785 is the rough equivalent of pi/4.

If an engine has a bore of 4 inches and a stroke of 3.48 inches, then the cylinder volume is figured: 0.785 × 4^2 × 3.48 = 43.708 cubic inches. Multiply this by the number of cylinders, which is 8, and you get 43.708 × 8 = 349.67, which is close to the engine displacement of the old small block Chevy 350 engine. The same approach can be used for engine sizes in metric. For example, a four-cylinder engine with a bore of 87 mm (8.7 cm) and a stroke of 99 mm (9.9 cm): 0.785 × 87^2 × 99 = 2,352 cubic centimeters or, as the engine is marketed, as a 2.4 L.

Given the formula above, calculate the engine displacements for the following:

1. 8-cylinder, bore 4 inches, and stroke 3.5 inches _____

2. 6-cylinder, bore 89 mm, and stroke 93 mm _____

3. 4-cylinder, bore 81 mm, and stroke 87.3 mm _____

F. English Measurement

English measurement uses the mile, yard, foot, inch, and fractions of the inch. This system has been in use for centuries but has been replaced in most parts of the world with the metric system. However, common uses for English measurement still occur, and you should be able to accurately read an English ruler and tape measure.

1. In Figure 4-6, measure from the starting line (A) to each lettered line and record the measurement in lowest terms. Note: Your measurement should reflect what each lettered line represents on an English ruler and not the actual distance from line A.

2. What is the smallest unit shown in Figure 4-6? _____

In the spaces provided, answer with the correct measurement in lowest terms for the distances in the figure below.

Example line = 4/16" or 1/4"

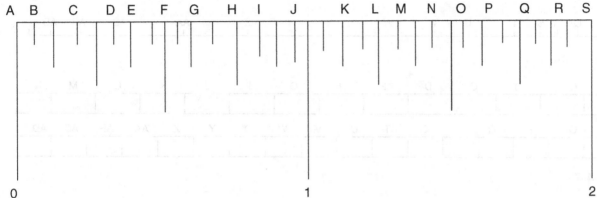

AB____   2. AC____   3. AD____   4. AE____   5. AF____   6. AG____

AH____   8. AI____   9. AJ____   10. AK____   11. AL____   12. AM____

AN____   14. AO____   15. AP____   11. AQ____   17. AR____   18. AS____

Figure 4-6

Using a tape measure, locate the following:

3. Measure and record the wheelbase of a vehicle in the lab. _____

4. Measure and record the distance between the two upright columns of an above-ground vehicle lift. _____

5. Measure the width of the lab garage door. _____

6. Measure the length of a wiper blade on a lab vehicle. _____

Using a ruler, measure and place a mark along the line at the given distance.

7. (1 inch) _____   8. (⅝ inch) _____

9. (¾ inch) _____   10. (1³⁄₁₆ inch) _____

Using an English ruler, measure along the lines in the upper part of Figure 4-7 and record the actual distances in the spaces provided at the bottom of the figure.

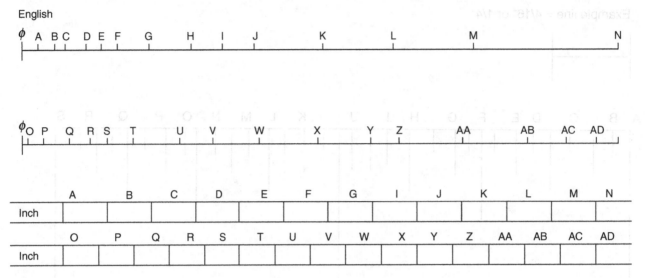

	A	B	C	D	E	F	G	I	J	K	L	M	N
Inch													

	O	P	Q	R	S	T	U	V	W	X	Y	Z	AA	AB	AC	AD
Inch																

Figure 4-7

G. Metric Measurement

Metric measurement is based on units of 10, which makes switching units very quick and easy since only the decimal place needs to be changed. The metric system uses the meter as its base unit. The meter is then divided into tenths, called decimeters. The decimeter is divided into tenths also, called centimeters. Centimeters are also divided into tenths, called millimeters. There are 10 decimeters, 100 centimeters, and 1,000 millimeters in a meter. 1,000 meters is a kilometer, which is roughly equivalent to 0.62 miles. A meter is roughly 39 inches long, making it slightly longer than a yard.

In Figure 4-8, use a metric ruler to measure from the starting line to each lettered line and record your readings in the spaces provided at the bottom of the figure.

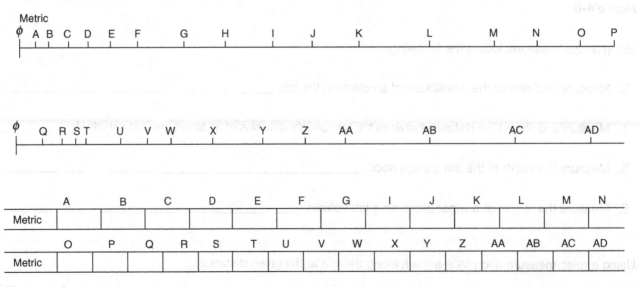

	A	B	C	D	E	F	G	I	J	K	L	M	N
Metric													

	O	P	Q	R	S	T	U	V	W	X	Y	Z	AA	AB	AC	AD
Metric																

Figure 4-8

Using a metric ruler, locate the following:

1. Length of a bolt provided by your instructor _____

2. Diameter of a fire extinguisher _____

3. Width of a lug nut provided by your instructor _____

4. Length of a wiper blade _____

5. (Instructor option) _____

Using a metric ruler, measure and place a mark along the line at the given distance from the left side of the line.

6. (10 mm) _____   7. (19 mm) _____

8. (13 mm) _____   9. (21 mm) _____

10. (17 mm) _____   11. (14 mm) _____

12. (22 mm) _____   13. (8 mm) _____

14. (12 mm) _____   15. (15 mm) _____

H. Using Measuring in Lab

Using a ruler and tape measure, obtain the following measurements:

1. Vehicle _____   Wheelbase _____

2. Vehicle _____   Wiper blades _____

3. Vehicle _____   Lug pattern _____

4. Vehicle _____   Generator pulley diameter _____

5. Vehicle _____   P/S pump pulley diameter _____

I. Labor Time

When a technician performs a job, he or she may be paid based on the labor time to perform the particular job. Labor times are found in estimating guides, also called *flat-rate guides*. These guides are used to provide an estimate of the amount of time required to perform tasks. For example, the labor time to replace a set of front brake pads may be one hour. If the technician replaced the pads in less than one hour, he or she could then move on to the next job while having been paid for an hour of work even though the actual work time was less. If, however, the technician takes one and a half hours to replace the brake pads, he or she will still be only paid for one hour of work, and so, lost money on the repair.

Labor guides break an hour down into tenths of an hour, or six minutes. Repair times are given in tenths of hours and are shown in decimal form. For example, a job that pays one half of an hour, labor is shown as 0.5 hours. Jobs may be anywhere from 0.1 hours to several hours depending on the difficulty of the work.

Using an estimating guide, find the labor times for the following items:

1. 2012 Chevy Malibu V6 front brake pad replacement _____

    _____

    _____

2. 2013 Honda Odyssey EX timing belt replacement _____

3. 2014 Ford F150 XLT 2WD lower ball joint replacement _____

4. 2012 Nissan Maxima spark plug replacement _____

5. 2008 Toyota Camry V6 front window motor replacement _____

J. Replacement Parts

Part of servicing and repairing cars and trucks is determining what parts need to be replaced, finding the parts cost and their availability, and preparing the estimate for the customer. Shops purchase parts at a price discount, often called *wholesale price*, and then mark the parts up and resell them to the customer. This is necessary for the shop to make money and enough profit to stay in business.

When preparing an estimate, you will often need to contact a local part supplier, which may be an aftermarket store such as NAPA, O'Reilly's, AutoZone, or similar store, or the parts may be purchased from a local new car dealership. Where the part is purchased depends on the type of part, the cost of the part, the availability, and the quality of the part. In some cases, the correct replacement part can only be bought from the dealer.

Using a school vehicle, your own vehicle, or one provided by your instructor, locate the following part information.

Locating Parts

Select a vehicle, either a school or your own (not a fantasy vehicle) for which to locate parts.

Year _____ Make _____ Model _____

Engine: _____ AT   MT   P/S   A/C   (circle all that apply)

You will need to access either OnDemand5 or AllData for the recommended fluids and then use the following websites, www.napaonline.com, www.autozone.com, or www.advanceautoparts.com to locate the following parts.

1. Engine oil filter Brand _____ Part number _____ Price _____.

2. Engine air filter Brand _____ Part number _____ Price _____.

3. Recommended engine oil weight _____ Brand _____ Price _____

4. Recommended coolant type _____ Brand _____ Price _____.

5. Wiper blade sizes _____ Brand _____ Price _____.

6. Recommended transmission fluid _____ Brand _____ Price _____.

7. Engine drive belt Brand _____ Part number _____ Price _____.

8. Spark plugs Brand _____ Part number _____ Price _____.

9. Battery Brand _____ Part number _____ Price _____.

Calculate the parts cost to perform an oil change, replace the oil filter, air filter, replace the wiper blades, drive belt, and spark plugs.

10. Total parts cost _____

K. Markup

Markup is the term used to describe the difference between the price paid for a part and the price at which the part is resold to the customer. For example, if the shop buys brake pads for $19 and sells to the customer for $38, the shop marked up the pads 200%, selling the pads for twice what the shop paid for them. The amount of markup on a part often varies based on the cost of the part and the manufacturer's suggested price for the part. Parts markup is necessary for the shop to be able to provide a warranty on the parts and labor. If the pads are noisy and the customer returns to have them replaced, the shop usually does not receive reimbursement from the part supplier to replace faulty parts. Therefore, the shop performs the work for free.

Determine your shop's cost and the list price for the following parts. Then, determine the markup for the parts for a late-model vehicle.

Year _____ Make _____ Model _____

1. Front brake pads   Cost $ _____   List $ _____   Markup $ _____

2. Water pump   Cost $ _____   List $ _____   Markup $ _____

3. Front shock/strut   Cost $ _____   List $ _____   Markup $ _____

4. Air filter   Cost $ _____   List $ _____   Markup $ _____

5. Starter   Cost $ _____   List $ _____   Markup $ _____

Next, determine the resale price of the following parts based on their cost and markup percentages given.

6. Timing belt: cost $29.78, markup 65 percent   Resale price $ _____

7. Radiator: cost $129.13, markup 35 percent.   Resale price $ _____

8. Brake pads: cost $33.49, markup 45 percent   Resale price $ _____

9. Lug nut: cost $1.25, markup 80 percent   Resale price $ _____

10. Upper control arm: cost $103.79, markup 55 percent   Resale price $ _____

L. Shop Supplies

Shop supplies are items such as shop rags, chemicals, and small fasteners or hardware that are used during most types of services. Some shops charge a separate fee for these items, often a percentage of the final repair cost up to a set limit. For example, many shops charge a 5% shop supply fee on each repair order up to a maximum of $5.

1. Make a list of items used in your shop that you think may be billed as shop supplies.

   _____

   _____

2. Based on the example of 5% up to $5, calculate the shop supplies charge for the following:

   a. Total repair bill $123.55 _____

   b. Total repair bill $75.89 _____

   c. Total repair bill $468.33 _____

M. Sales Tax

Sales tax is common in most states and localities and can vary from city to city and county to county. Sales tax can be applied to the total of parts and labor or only to parts or labor, depending on location. When writing estimates and completing repair orders, it is very important for you to correctly determine the sales tax so the customer is charged correctly.

Calculate the sales tax on the following:

1. Total repair bill $378.85, sales tax 6.75% _____

2. Total repair bill $39.90, sales tax 7.25% _____

3. Total repair bill $234.18, sales tax 5.75% _____

4. Determine the sales tax for your location. Tax rate _____

Based on your local sales tax, determine the amount of tax collected for the following:

5. Bill amount $88.89   Sales tax $ _____   Total $ _____

6. Bill amount $246.52   Sales tax $ _____   Total $ _____

7. Bill amount $512.13   Sales tax $ _____   Total $ _____

N. Balancing a Checking Account

Sooner or later you will have a checking or savings account. To keep track of how much money you have, you should periodically balance the account. This means you confirm how much you have in the account by comparing what the bank statement shows to what your own records show. This will help you track your income and expenses and hopefully keep you from overspending your account, which leads to fees and penalties.

To track your money, use a checking account register or a simple running record, like that shown in Figure 4-9. On the left side of the register, record the date and check number for transactions. In the center column, note what the transaction was, such as a deposit, cash withdrawal, debit, or similar. In the right columns, note the amount of the transaction and in the far right keep a running balance, as shown in Figure 4-9.

AD-Automatic Deposit • AP-Automatic Payment • ATM-Cash Withdrawal • DC-Debit Card • FT-Funds Transfer • SC-Service Charge • TD-Tax Deductible						
NUMBER OR CODE	DATE	TRANSACTION DESCRIPTION	PAYMENT, FEE, WITHDRAWAL(−)	✓	DEPOSIT, CREDIT (+)	$ BALANCE
	2/15	Paycheck deposit		✓	487|15	562|77
	2/16	Cash	50	✓		512.77
	2-18	Callahan Auto Parts	78.63	✓		434.14
	2-20	Car payment	288.60	✓		145.54
	2-20	gas	18	✓		127.54
	2/25	insurance	112.25			15.29
	3/1	deposit			390.41	405.70

Figure 4-9

When you receive your account statement, check the balance on the statement and compare it to the balance you show in your register. These will likely not be the same as you may have made many purchases and/or deposits since the statement was printed. To reconcile your balance with that shown in the bank statement, follow these steps:

1. Begin by checking off all transactions that have cleared the bank. This means on your register, check off all debits and deposits that show on the bank statement.

2. Add any uncleared debits to your account to your register balance. If you do not have any uncleared deposits, then your register balance plus uncleared debits should equal the balance shown on the statement.

3. If you have uncleared deposits, subtract the amount from the register balance. If the two numbers do not match, recheck that you have all transactions accounted for in your register and all cleared transactions on the bank statement.

4. If you still cannot find a reason for the numbers not to match, you may have a mistake in your register. Double check all of your addition and subtraction to check for mistakes.

If your balance and the balance shown by the bank still do not match, go back over every transaction during the statement dates. Make sure you recorded the correct amounts for both deposits and debits in your register. Be sure to add in any interest and subtract any fees, such as ATM fees or account maintenance charges.

O. Interest and Loans

At some point, you will likely start thinking about purchasing a vehicle or other expensive item, maybe a big toolbox, that will require taking out a loan. Borrowing money from a bank or a tool company means that you will make monthly payments and pay back more money than you borrowed. This extra money is called interest. Interest is the fee or charge you pay the bank for the privilege of using their money. How much interest you pay depends on many factors, including your credit rating (or lack or credit rating), the amount of money borrowed, how long the loan is for, and what the money is purchasing.

For example, maybe you decide to buy a new tool box, which costs $2,000. You do not have $2,000 in cash so you borrow money from the tool company for the box. The tool company decides to loan you $2,000 at 10 percent interest for one year. This is called a simple interest loan. The total amount of money you will pay can be easily calculated using the formula $I = Prt$. I equals the interest paid on the loan, P = the principal or the amount borrowed, r = the percentage rate of the loan, and t = the time in years. In this example, the interest equals to $200 since $(2,000)(0.10)(1) = 200$.

1. Based on this example, how much will your monthly toolbox loan payment be? _____

2. Based on the monthly payment, you decide to make the loan for two years instead of one, which will reduce your monthly payment. How much interest will you pay over a loan for two years? _____

3. How much will your toolbox payment be per month if you finance the loan for two years? _____

4. In this example, which loan would you prefer and why? _____
_____

IX. Science Skills

A. Energy Conversion

Many forms of energy are used in the automobile, and various energy conversions also take place. As discussed in Chapter 4, energy is not created or destroyed, it is used to perform some type of action. Following are types of energy conversions used in cars and trucks; match the energy type with its function.

Electrical to mechanical	Brake friction
Mechanical to electrical	Combustion
Chemical to thermal	Starter motor operation
Kinetic to heat	Generator operation
Thermal to mechanical	

B. Newton's Laws of Motion

Match the following examples with Newton's Laws of Motion.

Newton's First Law—Inertia

Newton's Second Law—Force = mass · acceleration

Newton's Third Law—Equal and opposite reaction

Weight transfer during braking _____

Accelerating from a stop _____

Brake caliper operation _____

Sliding to the outside of a turn while cornering _____

Increasing the weight of the vehicle but leaving the power output unchanged _____

C. Forces

To slow or stop the vehicle, all you have to do is press the brake pedal. But do you understand what takes place each time that pedal is pressed? The pedal itself is an application of one of the oldest forms of machines, the lever. Known since ancient times, the lever is not only one of the oldest machines, it is also one of the most simple. A lever applies force to gain mechanical advantage—for instance, when moving a heavy object. This is accomplished by increasing the distance over which the force is applied.

A teeter-totter is another example of a lever. If two people of the same weight sit the same distance apart from the middle, the teeter-totter will balance and both people will be suspended off the ground at equal heights. But what happens if one person weighs more than the other? To offset the imbalance, the heavier person must move forward, closer to the middle, or the lighter person must move farther away from the middle, to the very edge of the board, as shown in Figure 4-10.

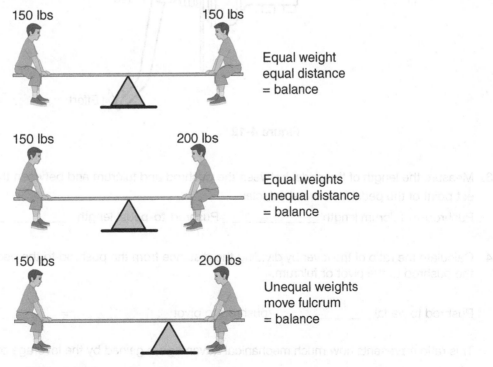

Figure 4-10

1. If the fulcrum stays in the middle, why must one person move to offset the imbalance?

   _____

2. If the two people stay in place, how must the fulcrum move to rebalance the teeter-totter?

   _____

     A teeter-totter is an example of a first-class lever. In a first-class lever, the fulcrum is located between the load and the effort, as shown in Figure 4-11. Using a shovel to dig up a plant is another example of a first-class lever in action.

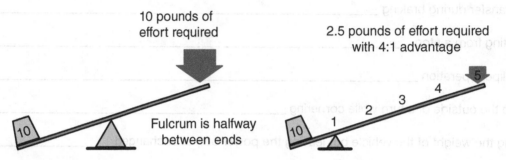

Figure 4-11

     The brake pedal in a car operates as a second-class lever. With a second-class lever, the fulcrum is at one end of the lever instead of the middle. The effort is applied to the other end of the lever, and the force is applied somewhere in between the effort and the fulcrum, as shown in Figure 4-12. If you look under the dashboard of a vehicle, you will be able to see how the brake pedal is mounted. Notice that the top of the pedal is the mounting point, which acts as the fulcrum. Below the fulcrum the pushrod is attached, which extends forward through the firewall to the brake booster and master cylinder. The footpad at the bottom of the pedal is where the effort is applied.

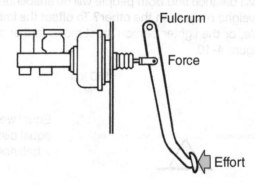

Figure 4-12

3. Measure the length of the pedal between the pushrod and fulcrum and between the pushrod and the lowest point of the pedal for several vehicles.

   Pushrod-to-fulcrum length _____ Pushrod-to-pedal length _____

4. Calculate the ratio of the lever by dividing the distance from the pushrod to the pedal by the distance from the pushrod to the pivot or fulcrum.

   Pushrod to pedal _____ /pushrod to pivot _____ = _____

   This ratio represents how much mechanical advantage is gained by the leverage of the brake pedal.

5. If the driver applies 50 pounds of force to the brake pedal, how much force is delivered by the brake pushrod? _____

6. If the driver applies 75 pounds of force to the brake pedal, how much force is delivered by the brake pushrod? _____

7. If the driver applies 150 pounds of force to the brake pedal, how much force is delivered by the brake pushrod? _____

8. In Figure 4-13, label the lever, pivot, and where the force is being applied.

Figure 4-13

A gear is a circular component that transmits rotational force to another gear or component. Gears have teeth so that they can mesh with other gears without slipping. The amount of leverage or mechanical advantage gained by the gear and ultimately, the gear ratio, is determined by the distance from the center of the gear to the end of a tooth, as shown in Figure 4-14.

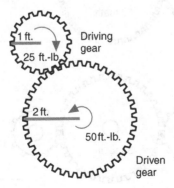

Figure 4-14

In Figure 4-14, the driving gear has a radius of 1 foot. As the driving gear turns clockwise, the driven gear rotates counterclockwise. The driven gear can apply twice the torque of the driving gear since the driven gear has a radius of 2 feet. By using a gear that is larger, more force can be applied since the distance from the center of the gear to the teeth creates leverage. This advantage allows gears to transmit a large amount of torque through the drivetrain to propel the vehicle.

Since the driven gear has twice as many teeth as the driving gear, it will turn at one-half the speed of the driving gear. The relationship between gear radius, and consequently, the number of gear teeth, is called *gear ratio*. The driving gear will turn twice for every one rotation of the driven gear, so the gear ratio is 2:1, shown in

Figure 4-15. If three gears are used, as shown in Figure 4-16, the center gear is an idler gear and does not affect gear ratio. The driving gear and driven gear teeth determine the gear ratio.

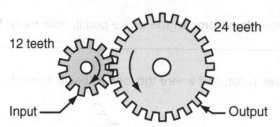

Gear ratio = 2 to 1
(2 revolutions input to
1 revolution output)

12 teeth   24 teeth

Input   Output

Figure 4-15

INPUT   IDLER   OUTPUT

Figure 4-16

9. Determine the gear ratios of the gears shown in Figure 4-17. Write your answers in the spaces provided in the figure.

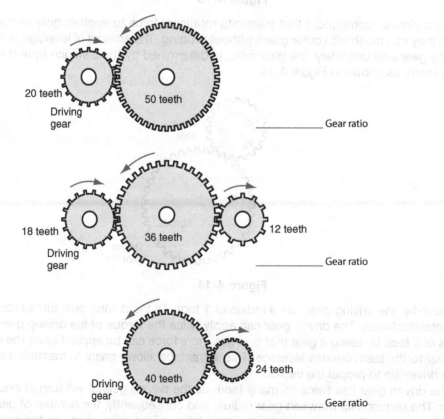

20 teeth
Driving gear
50 teeth
_____ Gear ratio

18 teeth
Driving gear
36 teeth
12 teeth
_____ Gear ratio

40 teeth
Driving gear
24 teeth
_____ Gear ratio

Figure 4-17

Pulleys, along with drive belts, are used to transmit motion. Located on the front of the engine, the crankshaft pulley is used to provide rotational force to other pulleys to drive accessories such as the generator, power steering pump, and water pump. An example of a front-end accessory drive arrangement is shown in Figure 4-18. You may notice that not all of the pulleys are the same size. Using pulleys of different diameters allows the various accessories to be driven at different speeds. For example, generator drive pulleys are typically about half the diameter of the crankshaft pulley. This means that for any given crankshaft rpm, the generator will be spinning about twice as fast.

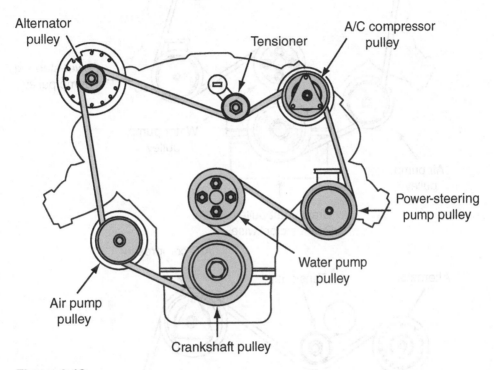

Figure 4-18

Because of the use of multi-rib serpentine belts in modern vehicles, either the rib or the flat side of the belt may drive pulleys. In addition some pulleys are driven opposite the engine. Most all newer engines rotate clockwise as seen from the front of the engine. However, some pulleys are driven counterclockwise based on how the drive belt is routed.

10. Examine the drive belt arrangements in Figure 4-19, and label the direction of rotation for each pulley.

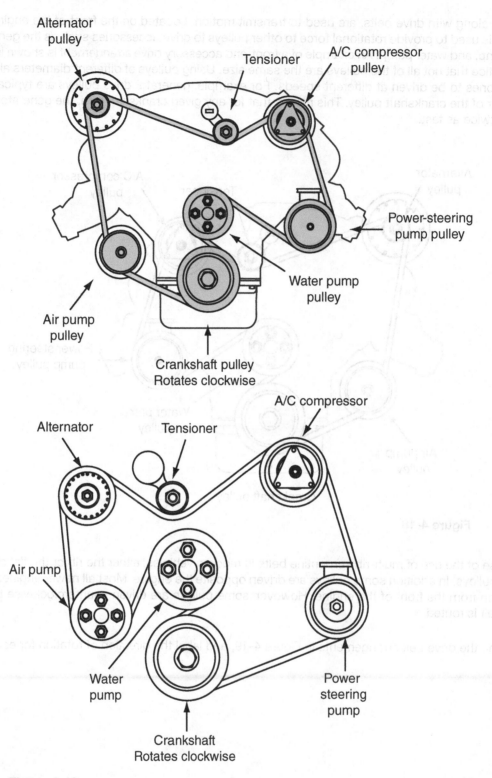

Alternator pulley

Tensioner

A/C compressor pulley

Power-steering pump pulley

Water pump pulley

Air pump pulley

Crankshaft pulley
Rotates clockwise

Alternator

Tensioner

A/C compressor

Air pump

Water pump

Power steering pump

Crankshaft
Rotates clockwise

Figure 4-19

D. Friction

In the automotive brake system, the pads and shoes create friction when applied; this in turn creates a lot of heat. Heat is a natural by-product of friction. Rub your hands together briskly and note how fast your skin warms.

1. Why does rubbing your hands together create heat? _____

2. Wet your hands with some water and rub them together briskly again. How did the water change the result?

_____

In automotive applications, there is friction used by the brake system, the friction of moving fluids such as motor oil and transmission fluid, friction between the vehicle and the air in which it is traveling, called drag, and friction between the tires and the ground.

Friction is created when uneven surfaces, in contact with each other, start to move relative to each other. Even though the two surfaces may appear smooth, there are slight imperfections that resist moving against each other. The amount of force required to move one object along another is called the coefficient of friction (CoF). Simply put, the CoF is equal to the ratio of force to move an object divided by the weight of the object, $CoF = F/M$. While several factors affect the CoF, such as temperature and speed, we will use a simple example of two bodies moving against each other. Imagine you have a 100 lb (45 kg) block of rubber on the floor of your lab. The force required to slide the block of rubber over concrete would be great. If it takes 100 lb. (45 kg) of force to slide the block, the CoF will be 100/100 or 1.0. Imagine the same block of rubber now sitting on the floor of an ice hockey rink. If it only takes 25 lb (11kg) of force to slide the block, what will the CoF be?

3. 25/100 = _____

To experiment with the CoF of various objects in your lab, you can make a small force gauge using an ordinary ballpoint pen, a rubber band, tape, and a paper clip. Assemble the parts as shown in Figure 4-20. Attach the paper clip to an object and try to pull it across a flat surface using the opposite end of your force meter. Measure the amount of extension of the pen tube from the body of the pen with a ruler. This will give you an idea of how much force is required to drag each object. Record your results below:

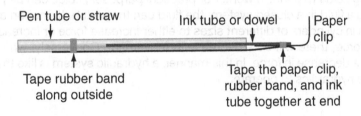

Pen tube or straw   Ink tube or dowel   Paper clip

Tape rubber band   Tape the paper clip,
along outside   rubber band, and ink
                          tube together at end

Figure 4-20

4. Object _____ Surface _____ Length of extension _____

5. Object _____ Surface _____ Length of extension _____

6. Object _____ Surface _____ Length of extension _____

7. Why did some objects require more force than others did? _____

8. What effect does the surface used to slide across affect the CoF? _____

9. If a liquid is placed between the two objects, what effect will that have on the CoF? _____

10. What could happen if the CoF of the brake pads or shoes was too high? _____

11. What could happen if the CoF of the brake pads or shoes was too low? _____

12. What factors do you think are involved in determining the correct CoF for a particular vehicle? _____

The CoF of brake linings can have an impact on how well a vehicle stops. Using lining materials with too high or low a CoF can cause brake performance issues, rapid wear, and customer dissatisfaction. Table 4-1 shows common brake material coefficients.

DOT Edge Code	Coefficient of Friction at 250° F and 600° F	Fade Probability
CC	0.0 to 0.15 both temps	
DD	0.15 to 0.25 both temps	
EE	0.25 to 0.35 both temps	0 to 25% at 600° F
FE	0.25 to 0.35 at 250° F temp 0.35 to 0.45 at 600° F	2% to 44% fade at 600° F
FF	0.35 to 0.45 at both temps	0 to 22% fade at 600° F
GG	0.45 to 0.55	Very rare
HH	0.55 to 0.65	Carbon/Carbon only—glows at about 3000° F
Organic linings	Cold: 0.44, warm: 0.48	
Semimetallic	Cold: 0.38, warm 0.40	
Metallic	Cold: 0.25, warm: 0.35	
Synthetic	Cold: 0.38, warm: 0.45	

E. Hydraulics

Hydraulics refers to using fluids to perform work. For practical purposes, fluids can be pressurized but not compressed. By pressurizing a fluid in a closed system, the fluid can transmit both motion and force. This is accomplished by using pistons in cylinders of different sizes to either increase force or increase movement. Whenever there is an increase in force, there will be a decrease in movement. Conversely, when there is an increase in movement, there will be a decrease in force. In this manner, a hydraulic system is like the lever. Locate two items around your lab that use hydraulics to operate.

1. Item one: _____

   How does this item use hydraulics? _____

2. Item two: _____

   How does this item use hydraulics? _____

A simple hydraulic system contains two equal-sized containers of a liquid with two equal-sized pistons. The two containers are connected with a hose or tubing. An example of this is shown in Figure 4-21. If one piston is pushed downward with 100 lb (45 kg) of force and moves down 10 inches, the piston in the second container will move upward 10 inches with the same 100 lb (45 kg) of force. Since the pistons are the same size, any force and movement imparted on one piston will cause the same reaction to the second piston.

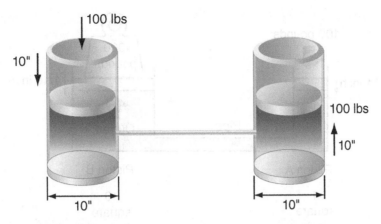

Figure 4-21

Where the use of hydraulics really provides advantage is when the sizes of the pistons are different; the resulting force and movement can be increased or decreased as needed. The pressure generated by the piston is a factor of the piston size. Input pressure is found by dividing force by piston size, or $P = F/A$. The smaller the input piston surface area is, the larger the force will be from that piston. The larger the input piston surface area is, the less the force will be from that piston, as shown in Figure 4-22. Conversely, the output piston force is proportional to the pressure against the surface area of the piston. An output piston that is larger than the input piston will move with greater force than the input but with less distance moved. To examine this principle, we will look at what is known as Pascal's principle or law.

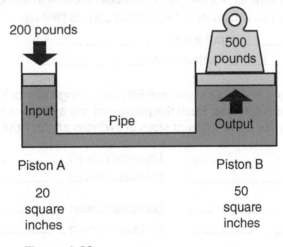

Figure 4-22

In Figure 4-23, the smaller piston on the left is our input piston and has a surface area of 1 square inch (6.45 cm²), and the larger piston is our output piston and has an area of 10 square inches (65.5 cm²). A force of 100 lb (45 kg) is exerted on the input piston, moving it downward 1 inch (2.54 cm). Using $P = F/A$, we can calculate $P = 100/1$, or 100 psi. Our output piston, which has a surface area of 10 square inches, will receive 100 pounds of pressure per square inch. This will result in an output force by piston 2 of 1,000 lb. However, because our output force has increased, our output movement will decrease. The larger piston will move one-tenth of the distance of the input piston; since the force was multiplied by 10, the distance will be divided by 10 as well. The 1 inch of movement of piston 1 turns into 1/10 of an inch of movement at piston 2. To calculate the forces and movement in a hydraulic circuit, use the following formulas:

$P_1 = F_1/A_1$ for piston 1 pressure

$F_2 = A_2/A_1 \cdot F_1$ for piston 2 force

$P_2 = F_2/A_2$ for piston 2 pressure

To practice using these principles, complete the following activity.

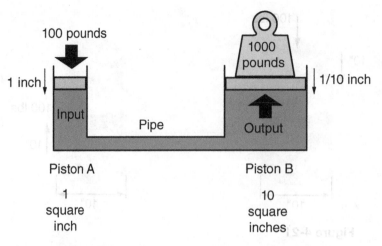

Figure 4-23

3. Calculate the forces and movements of a two-piston hydraulic circuit with an input piston of 2 square inches and an output piston of 10 square inches and an input force of 500 pounds (227 kg).

Input distance _____ Input force _____

Output distance _____ Output force _____

4. Calculate the forces and movements of a two-piston hydraulic circuit with an input piston of 2 square inches and an output piston of 1 square inch and an input force of 800 pounds (364 kg).

Input distance _____ Input force _____

Output distance _____ Output force _____

Obtain a selection of syringes from your instructor. Fill each syringe approximately halfway with water. Connect a hose between two different syringes. Push the plunger of one syringe, and note the reaction of the other. Perform this exercise several times using different sizes of syringes as the input and output pistons.

5. Syringe 1 diameter _____   Distance moved _____

    Syringe 2 diameter _____   Distance moved _____

6. Syringe 1 diameter _____   Distance moved _____

    Syringe 2 diameter _____   Distance moved _____

7. Syringe 1 diameter _____   Distance moved _____

    Syringe 2 diameter _____   Distance moved _____

8. Syringe 1 diameter _____   Distance moved _____

    Syringe 2 diameter _____   Distance moved _____

F. Electricity

Mostly everyone is familiar with two very common electrical occurrences, lightning, and static electricity. Lightning is an electrical discharge due to static electricity, just like when you get shocked touching a doorknob on a dry winter day. The difference is the energy in a lightning bolt is many, many times greater.

1. Describe what you think are the conditions necessary for getting a static electricity shock. _____
   _____

2. Why do you think a lightning discharge is so much more powerful than a static electricity shock? _____
   _____

   The term *static electricity* refers to a buildup of electrical charges where there is typically poor electrical conductivity. When there is an electrical charge, there is the potential for electron flow. When an electron leaves an atom, it leaves behind an opening. The atom from which the electron left will now have a positive charge due to the loss of the negatively charged electron. The atom the electron moved to now has a negative charge. Whenever an imbalance occurs, there is the potential for electrical current flow. When an electrical force is present, and electrons move from atom to atom, electrical current flow exists. If a material does not easily accept the movement of electrons, it is an insulator or a poor conductor. Metals such as copper, silver, gold, and aluminum can readily give up and transfer electrons, and are considered good electrical conductors. Figure 4-24 shows how electrons move between atoms.

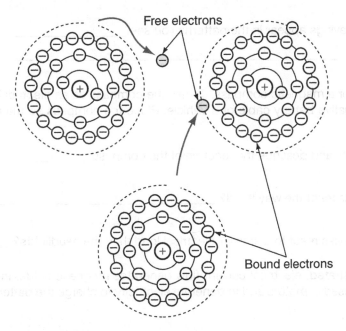

Figure 4-24

   If during this static electricity transfer, dissimilar charges are present, the charges will attract. If the charges are similar, the charges will repel. Just as two magnets will repel when north poles are placed close together and attract if the north and south poles are near each other. If you have ever pulled a wool sweater on over your head and had your hair stand on end, you have experienced this phenomenon. When you pulled the sweater over your head, electrons transferred from the wool to your hair. Similarly, rubbing a balloon against a sweater can cause enough of a static charge to allow the balloon to "stick" to a wall.

3. Inflate two balloons and attach a length of string to each knotted end. Rub one balloon against a piece of clothing. Then hold the string so that the balloons hang down near each other. Describe what takes place.
   _____

4. Rub the balloons against your clothing and attempt to stick them to various surfaces. What surfaces stuck the best? _____ Which were the worst? _____

5. Describe why you think some surfaces hold the balloons better or worse than others. _____

6. What environmental factors affect static electricity? _____

When discussing electricity, it is necessary to include magnetism because the two are interrelated and inseparable. Electromagnetism refers to the interaction of electricity and magnetism. When a current flows through a conductor, a magnetic field develops around the conductor. Likewise, when a conductor is moved through a magnetic field, current is induced into the conductor. This means that electricity can be generated by moving a conductor through a magnetic field. It is this principle that is the basis of electrical power generation, whether under the hood of a car with a generator or at a power generation facility, like a hydroelectric dam or a nuclear power plant. Electromagnetism also defines how relays, electric motors, and ignition coils operate.

Obtain a magnet and some iron shavings from your instructor. Place a piece of paper on top of the magnet, then sprinkle some of the shavings on the paper over the magnet.

7. What shapes do the shavings show on the paper? _____

8. Describe why the shavings appear in the patterns you see. _____

_____

Ask your instructor for a magnetic compass. To use the compass as an electrical current detector, place it along the positive or negative battery cable of a vehicle. Place the compass so the needle is parallel with the cable.

9. Turn on the headlights and describe the reaction of the compass. _____

10. Why did the compass react the way it did? _____

11. How would the compass react to a smaller electrical load than the headlights? _____

As you have demonstrated, electrical current flow generates magnetic fields. In Chapter 17, you will see how magnetic fields are used in motors and to generate electricity to charge the battery and power the vehicle's electrical system.

G. Pressure and Vacuum

Air pressure, or atmospheric pressure, is a measurement of the weight of the air from the edge of the atmosphere down to ground level. We do not often think about the air that surrounds us, but the weight of that air has a huge impact on our lives. Air pressure at sea level is about 14.7 psi. This means that if we could take a 1-square-inch column of air from the ground up to the edge of space, it would have a weight of approximately 14.7 pounds, as shown in Figure 4-25. As you move higher and higher above sea level, air pressure decreases.

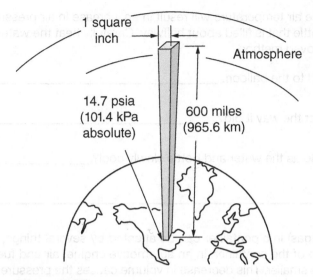

Figure 4-25

1. Explain why air pressure decreases with altitude. _____

Since air pressure is a product of our atmosphere, it is subject to changes in the atmosphere. In addition, since air pressure is affected by the weather, it can be easily measured and recorded in your lab on a daily basis by making this simple barometer. Take a glass container and stretch a balloon or similar material over the opening of the jar. If necessary, use a rubber band to seal the balloon around the jar: it must be airtight. Next, attach a straw horizontally to the balloon with a piece of tape. The height of the end of the straw can be marked on a piece of paper to record the changes in air pressure. As the local air pressure changes, the straw will rise or drop with the pressure changes. Record the height over several days or weeks to track the changes in air pressure.

2. Describe how this simple barometer operates. _____

3. What causes the up or down movement of the straw? _____

4. If the air in the jar were heated or cooled, how would the straw react? _____

5. How does temperature affect the pressure inside the jar? _____

We saw in the last activity that the air pressure changes, but what can cause the pressure to change? Temperature is one factor that can cause pressure change. When air is heated, it rises, and when air is cooled, it falls. The reason for this is that as air is heated, it weighs less per unit of volume than cooled air. Since the hotter air is less dense, the cooler, denser air lifts it. To demonstrate how temperature affects pressure, fill a plastic milk jug approximately one-third of the way full with very hot water. Use caution when working with hot water to avoid being accidentally burned. Screw the cap on securely, and wait about an hour.

6. What happened to the jug? _____

7. Why did the jug react in the way it did? _____

8. How did the outside air pressure affect the jug? _____

To see how increasing the air temperature will result in an increase in air pressure, place a balloon over the opening of a plastic water bottle that is filled about halfway. Carefully heat the water bottle with a heat gun, hair dryer, or other safe and approved method.

9. Describe what happened to the balloon. _____

10. Why did the balloon react the way it did? _____

11. What does the balloon do as the water and bottle slowly cool? _____

12. Why did this happen? _____

The pressure of air (or a gas) in a container can be affected by several things, such as the temperature in the container and the volume of the container. In an automotive engine, air and fuel is drawn into the cylinder, which once sealed, becomes smaller. This decrease in volume causes the pressure and temperature of the gas to increase as the molecules are forced more tightly together.

When the volume of the cylinder is increased, such as when the piston moves from TDC to BDC on the intake stroke, the pressure in the cylinder decreases.

13. Why does the pressure decrease? _____

14. What causes the air to enter the cylinder? _____

Place the small open end of a syringe against one of your fingers and draw the plunger out until it stops moving.

15. Describe what happens. _____

16. What was the effect within the syringe by pulling out on the plunger? _____

Attach a pressure/vacuum gauge to a syringe with the plunger about halfway pulled out. Pull the plunger back to the end of its travel and note the reading on the gauge.

17. Gauge reading _____

18. With the plunger at the end of its travel, push the plunger in and note the reading on the gauge. _____

Obtain two syringes of equal size and pull each plunger out about halfway. Then connect the two together with a piece of vacuum hose.

19. Push on one plunger and note what happens to the other plunger. _____

20. Was the result what you expected? Why or why not? _____

21. What caused the movement of the second plunger? _____

22. Now, pull on the plunger and note what happens. _____

23. Why did the second plunger move? _____

24. How did the change in pressure in the first syringe affect the second syringe? _____
Obtain a vacuum pump and a cone-shaped dispensing lid from a differential fluid bottle from your instructor. Next, attach the cap to an empty motor oil quart bottle. Pull vacuum on the bottle via the cap and watch the oil bottle.

25. What happens to the oil bottle? _____

26. Why did the bottle react the way it did? _____

27. Release the vacuum and note the effect on the bottle. Explain the result. _____

Using a vacuum pump and a vacuum chamber connected to a bottle of oil, apply vacuum and watch the oil.

28. Why did the oil move from the oil bottle? _____

You have now demonstrated a basic physical principle; that pressure moves from a higher pressure to a lower pressure.
Next, place a balloon into an empty water bottle and leave the balloon's opening above the opening of the bottle. Now try to inflate a balloon that is inside the bottle.

29. Explain what happened. _____

30. Why did the balloon not inflate normally? _____

31. What could be done to allow the balloon to properly inflate? _____

The balloon was trying to inflate within the bottle and was exerting pressure against the air already inside the bottle. Since the air in the bottle had nowhere to go, the pressure equalized against the pressure you tried to place in the balloon, causing a pressure balance. The harder you tried to blow up the balloon, the more the pressure in the bottle worked against you. Now place a small hole in the bottom of the bottle and repeat the experiment.

32. Describe what happened. _____

33. Why was the balloon able to inflate more than before? _____

34. If you plug the hole in the bottle and release the air from the balloon, what will happen to the bottle? _____

_____

35. Why? _____

You have seen how pressure can move from high to low and how to increase pressure. Next, we will examine low pressure, also called vacuum. Vacuum is pressure that is less than atmospheric pressure. Pressure can be measured with a pressure gauge, such as a tire pressure gauge, but a tire gauge is calibrated to read atmospheric pressure as zero. This is referred to as psi gauge, or psig. A flat tire will read zero psi on a pressure gauge, but it actually has 14.7 psi (or equivalent local) pressure. To read pressures lower than atmospheric, you will need either a gauge that reads psi absolute (psia) or a vacuum gauge. Using a vacuum gauge and a pressure chart, you can easily see what a vacuum gauge reading reflects in actual air pressure. Connect a vacuum gauge to a vacuum hose and start the engine.

36. Record the gauge reading. _____

37. What does this equate to in actual air pressure? _____

38. Why is the engine producing vacuum? _____

39. What does the vacuum gauge read if the engine speed is increased? _____

H. Sound

Sound is a result of vibrations, though some vibrations can occur at frequencies either below or above the human range of hearing. As an object vibrates, it moves the air around the object, causing the air itself to vibrate. These vibrations are detected by the bones in our inner ears, producing the sounds that we hear. Frequency is the number of times a vibration occurs and is usually rated in Hertz (Hz), which is the number of cycles per second, shown in Figure 4-26. A sound with a vibration of 200 Hz will have 200 oscillations per second. A sound with a frequency of 2 KHz (2,000 Hz) will oscillate 2,000 times per second. The lower the Hertz rate is, the lower will be the frequency and the tone of the sound. The higher the frequency is, the higher pitch of the sound will be. Humans can usually hear sounds between 20 Hz and 20,000 Hz. As an example, the low string on a bass guitar vibrates at about 41 Hz, and the low string on a six-string guitar vibrates twice as fast at 82 Hz. The highest note on a six-string guitar vibrates around 1,300 Hz.

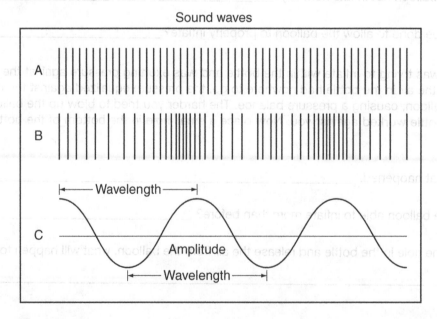

Figure 4-26

Being able to identify various sounds is an important diagnostic skill. By listening to a noise and its relation to engine speed or wheel speed, you can more easily locate the source of the problem.

1. A customer complains there is a high-pitched rattling noise from the vehicle that gets louder and increases in frequency when engine speed is increased. The noise occurs in Park and while driving. Which of these is the most likely cause and why?

    a. Loose wheel bearings

    b. Loose drive belt

    c. Loose exhaust shield

    d. Loose motor mount

    Why? _____

2. A customer states that a low-sounding rumble or roar occurs as the car is moving. The noise increases in pitch (frequency) as speed increases. Which of these is the most likely cause and why?

    a. Worn wheel bearing(s)

    b. Worn drive belt

    c. Broken exhaust shield

    d. Loose motor mount

    Why? _____

3. A customer brings his or her car in for a high-pitched squealing sound when the car is moving. The noise changes slightly when the brakes are applied and stops when the car stops moving. Which of these is the most likely cause and why?

    a. Worn wheel bearing(s)

    b. Brake pad wear indicator

    c. Broken exhaust pipe

    d. Loose drive belt

    Why? _____

X. Basic Chemistry

1. List two examples of chemical reactions that take place in the automobile.

    _____

    _____

2. Based on one of the examples you listed above, describe in detail what happens during the process, including what causes the reaction to take place, what factors affect the reaction, and what changes take place from the start to the finish of the reaction.

    _____

    _____

    _____

    _____

Lab Worksheet 4-1

Name _____ Station _____ Date _____

Communication Skills

ASE Education Foundation Correlation

This lab worksheet addresses the following **RST** task:

6-7. Communicates (written and verbal) effectively with customers and coworkers.

Description of Vehicle

Year _____ Make _____ Model _____

VIN: _____ Engine _____

Procedure

Using the following worksheet, contact a local parts supplier and obtain the following information necessary to complete an estimate.

1. Name of parts store and location   _____

2. Name of parts person taking the call _____

3. Use the following script to find the cost and availability of the parts:

   "Hello, this is (your name here) calling from the automotive program at (school name here). I need our cost on the following parts and their availability."

   a. Part _____   Cost _____

   b. Part _____   Cost _____

   c. Part _____   Cost _____

   d. Part _____   Cost _____

   e. Availability of the parts _____

4. Record the cost and availability in the spaces above. Once you have the parts cost and their availability, thank the person on the phone and tell him or her that you will call back once you find out if you will need the parts.

Instructor's check _____

CHAPTER 5

Wheels, Tires, and Wheel Bearings

Review Questions

1. Tires _____ the weight of the vehicle, absorb much of the road _____, and are vital to how the vehicle handles and how well it _____.

2. Tires that rely on air pressure are called:
   a. hydraulic.
   b. pneumatic.
   c. aquatic.
   d. electronic.

3. True or False: With the weight of the vehicle on the tire, the tire will remain perfectly round.

4. A tire contact patch of 36 square inches at 28 psi can support how much weight?
   a. 960 pounds
   b. 1255 pounds
   c. 1008 pounds
   d. 888 pounds

5. Explain why temporary spare tires require higher air pressure than the standard size tires?

   _____

   _____

6. When the weight of the vehicle is on the tires, the tires tend to _____ slightly.

7. When the temperature of the tire increases, such as during driving, the air pressure inside the tire _____.

8. The type of tire balance shown in Figure 5-1 is:

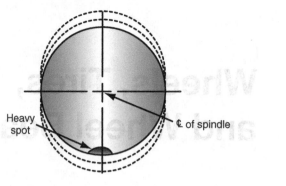

Heavy
spot

₵ of spindle

Figure 5-1

   a. dynamic.
   b. lateral.
   c. radial.
   d. static.

9. Dynamic wheel imbalance causes the steering wheel to shake or _____.

10. *Technician A* says radial tires are the most common tire type used today. *Technician B* says radial tires use layers of materials such as steel and polyester in their construction. Who is correct?
   a. Technician A
   b. Technician B
   c. Both a and b
   d. Neither a nor b

11. In general, the shorter a tire's sidewall is, the _____ the tire will be.

12. Identify the tire types shown in Figure 5-2 (a through c).

a. _____

b. _____

c. _____

Figure 5-2

13. List four different types of tires.

_____

_____

14. The majority of tires used on passenger vehicles are _____ season tires.

15. Sports car tires may not perform well in _____ temperature operation.

16. An _____ tire is one that has different tread patterns on the inside and outside sections of the tread.

17. A _____ tire is one that has a tread pattern that is made to rotate primarily in one direction.

18. Most temporary spare tires require an inflation pressure of _____ psi.

19. Self-supporting run-flat tires require the use of a tire _____ monitoring system.

20. Which of the following is not part of the tire size marking on a tire?
    a. Rim diameter
    b. Rim width
    c. Tread section width
    d. Aspect ratio

21. The _____ and _____ ratings use letters, from C to A or AA to indicate overall tire quality.

22. Most passenger car tires have maximum inflation pressures between _____ psi and _____ psi.

23. The last four digits of the _____ number provide the production date of the tire.

24. List three types of tire problems that can result from tire plies that are not molded correctly.

    _____

    _____

25. Tire pull can be caused by _____ meaning the tire is slightly cone shaped.

26. Steel wheels are often covered with _____ to enhance the look of the vehicle.

27. A hub-centric wheel means that the wheel is centered to the vehicle by the
    a. lug nuts.
    b. hub.
    c. hub cap.
    d. lug nuts, hub, and hub cap.

28. *Technician A* says changing wheel offset can affect wheel bearing life. *Technician B* says changing wheel offset can cause the tire to interfere with the brake system. Who is correct?
    a. Technician A
    b. Technician B
    c. Both a and b
    d. Neither a nor b

29. *Technician A* says overtightening lug nuts can damage the hub. *Technician B* says overtightening the lug nuts can damage the brake rotors. Who is correct?

    a. Technician A

    b. Technician B

    c. Both a and b

    d. Neither a nor b

30. The two types of tire pressure monitoring systems (TPMS) are

    a. static and dynamic.

    b. direct and indirect.

    c. lateral and radial.

    d. internal and external.

31. Explain the differences between the two types of tire pressure monitoring systems.

    _____

    _____

32. Describe why it is important to use caution when inspecting a tire.

    _____

    _____

33. Explain how to use a tire machine to dismount and remount a tire on a wheel.

    _____

    _____

    _____

    _____

34. Tire pressure should be set when the tire is _____. If tire pressure is checked when the tire is hot, the pressure may read _____.

35. Draw the two common patterns for performing a tire rotation.

36. Explain what is indicated by finding rubber dust inside of the tire.

_____

_____

37. If a wheel and tire are stuck on the hub, describe what steps should be taken to break the wheel loose from the hub.

_____

_____

38. Which of following is not a common reason for tire air loss?

    a. Debris punctured the tread.

    b. Rust or corrosion around the rim and bead.

    c. Cracks in the rim.

    d. Leaking valve stem or core.

39. A customer complains that the vehicle's steering wheel shakes while driving on the freeway but is fine at lower speeds. What is the most likely cause?

    a. Loose wheel bearings

    b. Tire imbalance

    c. Tire conicity

    d. Low tire pressure

40. A wheel/tire that is _____ out of balance will try to bounce or hop as it is driven.

41. *Technician A* says excessive wheel/tire runout can cause a vibration. *Technician B* says a bent hub can cause a vibration. Who is correct?

    a. Technician A

    b. Technician B

    c. Both a and b

    d. Neither a nor b

42. A vehicle has an illuminated TPMS light on the dash. *Technician A* says one or more tires may have low air pressure. *Technician B* says if the TPMS light stays on with the tires inflated correctly it means that there is a fault in the system. Who is correct?

    a. Technician A

    b. Technician B

    c. Both a and b

    d. Neither a nor b

43. Explain how bearings, particularly ball and roller bearings, reduce friction.

_____

_____

44. *Technician A* says all FWD vehicles use a sealed wheel bearing and hub assembly that is replaced as a unit. *Technician B* says some FWD vehicles use tapered roller bearings for the front wheel bearings. Who is correct?
    a. Technician A
    b. Technician B
    c. Both a and b
    d. Neither a nor b

Activities

1. For a vehicle specified by the instructor, use a ruler to determine the approximate contact patch area for a tire sitting on the floor and at its proper inflation pressure.

   a. Contact patch width _____ Length _____ Total area _____

   b. Multiply the tire pressure by the contact patch area to determine how much weight can be carried. _____ lb (kg)

   c. Record the maximum load rating specified on the tire sidewall. _____

   d. Find the difference between the specified maximum load and the weight you determined in question b. _____

   e. If there is a difference in the two numbers, why? _____

   f. How does inflation pressure affect the load carrying capacity of the tire? _____

   g. How does tire size (width) affect tire load carrying capacity? _____

   h. Determine if the vehicle has a temporary spare tire; if so, what is the recommended pressure for this tire? _____

   i. Why is the tire pressure for a mini-spare higher than the pressure for the regular tires?

   _____

2. Tire and wheel construction

   a. Label the parts of the tire shown in Figure 5-3 (A through C).

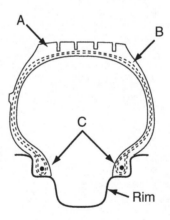

Figure 5-3

b. Label the tire dimensions in Figure 5-4.

P	215	65	R	15	89	H

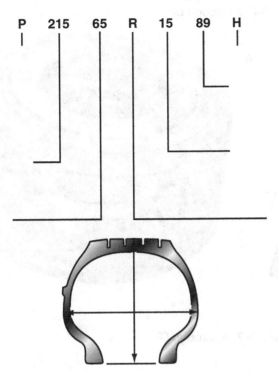

Figure 5-4

c. Label the parts of the wheel shown in Figure 5-5 (A through H).

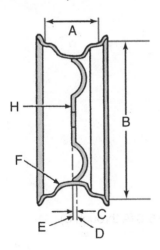

Figure 5-5

d. Match the tire sidewall information with its description. Passenger car tire

Section width	93
Aspect ratio	P
Rim diameter	18
DOT number	235
Speed rating	40
Load index	MXTB A9A 1511

3. Wheel bearing

   a. Label the types of wheel bearings shown in Figure 5-6 (a through c).

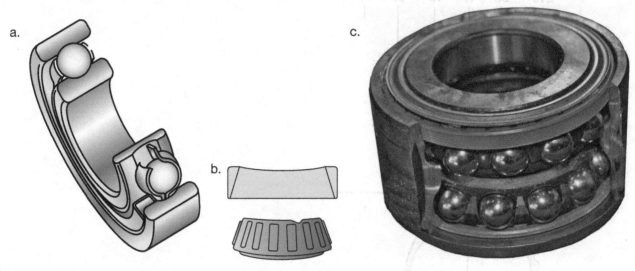

a.

b.

c.

Figure 5-6

   b. Label the parts of the wheel bearing shown in Figure 5-7 (A through C).

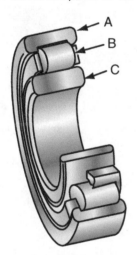

A
B
C

Figure 5-7

   c. Label the forces that act upon wheel bearings during operation in Figure 5-8 (A through C).

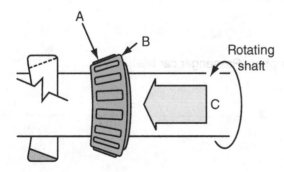

A
B
Rotating shaft
C

Figure 5-8

Lab Worksheet 5-1

Name _____ Station _____ Date _____

Inspect FWD Wheel Bearings

ASE Education Foundation Correlation

This lab worksheet addresses the following **MLR** task:

3.D.1. Inspect, remove, and/or replace bearings, hubs, and seals. **(P-2)**

Description of Vehicle

Year _____ Make _____ Model _____

Engine _____ AT MT CVT PSD (circle which applies)

Procedure

1. Describe the common symptoms of a faulty wheel bearing. _____
   _____
   _____

2. Does the bearing make noise when driving the vehicle?   Yes _____   No _____

3. If the bearing is making noise, does the noise change when turning a corner?

   Yes _____   No _____

   If yes, describe how the noise changes and what this indicates. _____
   _____

4. Raise and support the vehicle and spin the tire by hand. Note any roughness or looseness in the bearing.
   Note your findings. _____
   _____

5. Grasp the tire at the three o'clock and nine o'clock positions and shake the tire. Is any looseness felt?

   Yes _____   No _____

   If yes, is the movement in the wheel bearing?   Yes _____   No _____
   _____

6. Based on your inspection, what is the recommended action? _____
_____

7. Remove any tools and equipment and clean up the work area.

Instructor's check _____

Lab Worksheet 5-2

Name _____ Station _____ Date _____

Replace FWD Wheel Bearings

ASE Education Foundation Correlation

This lab worksheet addresses the following **MLR** task:

3.D.1. Inspect, remove, and/or replace bearings, hubs, and seals. **(P-2)**

Description of Vehicle

Year _____ Make _____ Model _____

Engine _____AT MT CVT PSD (circle which applies)

Procedure

1. Refer to the manufacturer's service procedures to replace a front wheel bearing. Summarize the procedure.

   _____

   _____

   _____

2. Safely raise and support the vehicle. Remove the wheel and tire on the side of the vehicle being serviced.

   Instructor's check _____

   The following procedures are general for most FWD vehicles. Refer to the manufacturer's service information if available.

3. Loosen and remove the drive axle nut.

4. Separate the steering knuckle from either the lower control arm or strut.

   Instructor's check _____

5. Remove the outer CV joint from the wheel bearing. Be careful not to damage the CV shaft, boots, or joints.

   Instructor's check _____

6. Remove the bolts securing the wheel bearing to the steering knuckle and remove the wheel bearing.

   Instructor's check _____

7. Install the new wheel bearing and torque the bolts to specifications.

    Torque specifications. _____

8. Reinstall the outer CV shaft into the wheel bearing. Reattach the steering knuckle to the control arm or strut. Torque all fasteners to specifications.

    Torque specifications_____

9. Install and torque a new axle nut to specifications. Do not use power tools to seat the nut before torquing. Overtightening the axle nut can damage the new bearing and cause premature failure.

    Torque specifications_____

    Instructor's check _____

10. Reinstall the wheel and tire. Torque the wheel lugs to specifications.

    Torque specifications_____

11. Lower the vehicle back on the floor.

12. Remove any tools and equipment and clean up the work area.

    Instructor's check _____

Lab Worksheet 5-3

Name _____ Station _____ Date _____

Replace Studs

ASE Education Foundation Correlation

This lab worksheet addresses the following **MLR** task:

3.E.4. Inspect and replace drive axle wheel studs. **(P-1)**

Description of Vehicle

Year _____ Make _____ Model _____

Engine _____ AT MT CVT PSD (circle which applies)

Procedure

Safely raise and support the vehicle.

1. Begin your inspection by checking all of the lug nuts and studs for signs of damage, for missing lug nuts and studs, and if all lug nuts and studs are of the same type and size. Summarize your findings.

_____

_____

_____

2. Select the appropriate size socket(s) and loosen all of the lug nuts and studs. Note if any of the lug nuts and studs require more or less effort than others as this can indicate improper torque and damage to the lug nuts or studs. Note your findings.

_____

_____

_____

3. Remove the lug nuts and studs and set them aside for inspection. Remove each wheel and tire assembly. Inspect all of the lug nuts and studs for signs of damage. Note your findings. _____

_____

_____

_____

4. If the lug nut studs have damaged threads, the stud may be repairable with a thread file or by using a die to clean the threads. Before attempting to repair the threads, you need to determine the stud thread pitch and diameter.

   Thread pitch _____

   Diameter _____

5. Using a thread file with the correct thread pitch, attempt to straighten the damaged threads. Be sure to keep the file properly aligned with the stud to prevent damaging the threads.

   Instructor's check _____

6. If the threads require using a die, ensure you have the correct thread pitch and diameter. Using the wrong size die can cause irreversible damage to the stud.

   Die used _____

7. Carefully thread the die onto the stud. Turn the die approximately one-quarter to one-half turn and then back it off slightly. Continue down the length of the stud until the threads are repaired.

   Instructor's check _____

If the threads are badly damaged or the stud is broken, you will have to replace the stud. Refer to the manufacturer's service information for the correct procedure.

8. Remove the damaged wheel stud from the axle hub.

   Instructor's check _____

9. Compare the replacement stud to the original. Ensure it is the correct length, diameter, and thread pitch.

10. Insert the stud into the axle hub. There are several methods to fully seat the wheel stud. Describe the method used.

    _____

    _____

11. Thread a lug nut onto the new stud and verify proper operation.

    Instructor's check _____

12. Once finished, put away all tools and equipment and clean up the work area.

    Instructor's check _____

Lab Worksheet 5-4

Name _____ Station _____ Date _____

Tire Size Information

ASE Education Foundation Correlation

This lab worksheet addresses the following **MLR** task:

4.D.1. Inspect tire condition; identify tire wear patterns; check for correct tire size, application (load and speed ratings), and air pressure as listed on the tire information placard/label. **(P-1)**

Procedure

Locate the following tire information:

1. Vehicle year _____   Vehicle make and model _____

   Tire manufacturer _____   Tire model _____

   Tire size _____   Treadwear rating _____

   Traction rating _____   Temperature rating _____

   Load rating _____   Speed rating _____

Locate the tire decal on this vehicle and determine if the installed tire is correct according to the decal.

   Correct   Yes _____   No _____

   If not, what is incorrect? _____

2. Vehicle year _____   Vehicle make and model _____

   Tire manufacturer _____   Tire model _____

   Tire size _____   Treadwear rating _____

   Traction rating _____   Temperature rating _____

   Load rating _____   Speed rating _____

Locate the tire decal on this vehicle and determine if the installed tire is correct according to the decal.

   Correct   Yes _____   No _____

   If not, what is incorrect? _____

Lab Worksheet 5-5

Name _____ Station _____ Date _____

Tire Pressure

ASE Education Foundation Correlation

This lab worksheet addresses the following **MLR** task:

4.D.1. Inspect tire condition; identify tire wear patterns; check for correct tire size, application (load and speed ratings), and air pressure as listed on the tire information placard/label. **(P-1)**

Procedure

Vehicle year _____   Vehicle make and model _____

Tire pressure spec _____   Spare pressure spec _____

RF tire pressure _____   LF tire pressure _____

RR tire pressure _____   LR tire pressure _____

Vehicle year _____   Vehicle make and model _____

Tire pressure spec _____   Spare pressure spec _____

RF tire pressure _____   LF tire pressure _____

RR tire pressure _____   LR tire pressure _____

Vehicle year _____   Vehicle make and model _____

Tire pressure spec _____   Spare pressure spec _____

RF tire pressure _____   LF tire pressure _____

RR tire pressure _____   LR tire pressure _____

Remove any tools and equipment and clean up the work area.

Instructor's check _____

Lab Worksheet 5-6

Name _____ Station _____ Date _____

Tire Wear

ASE Education Foundation Correlation

This lab worksheet addresses the following **MLR** task:

4.D.1. Inspect tire condition; identify tire wear patterns; check for correct size and application (load and speed ratings) and air pressure as listed on the tire information placard/label. **(P-1)**

Description of Vehicle

Determine any wear concerns for the tires specified by your instructor.

Vehicle year _____   Vehicle make and model _____

Procedure

1. Inspect the tires for uneven or excessive wear. Note your findings for each tire.

   RF tire wear _____   LF tire wear _____

   RR tire wear _____   LR tire wear _____

2. Using a tread depth gauge, measure the tread depth, and record it for each tire.

   RF depth _____   LF depth _____
   RR depth _____   LR depth _____

   Vehicle year _____   Vehicle make and model _____

3. Inspect the tires for uneven or excessive wear. Note your findings for each tire.

   RF tire wear _____   LF tire wear _____

   RR tire wear _____   LR tire wear _____

4. Using a tread depth gauge, measure the tread depth and record it for each tire.

   RF depth _____   LF depth _____

   RR depth _____   LR depth _____

Lab Worksheet 5-7

Name _____ Station _____ Date _____

Rotate Tires

ASE Education Foundation Correlation

This lab worksheet addresses the following **MLR** task:

4.D.2. Rotate tires according to manufacturer's recommendations including vehicles equipped with tire pressure monitoring systems (TPMS). **(P-1)**

Description of Vehicle

Year _____ Make _____ Model _____

Engine _____ AT MT CVT PSD (circle which applies)

Procedure

1. Find the vehicle lift points and support correctly.

   Instructor's check _____

2. Locate manufacturer-recommended tire rotation pattern, and include a diagram of pattern, or draw and describe.

3. Check all wheels and tires. Note any problems.

   Wheel damage, cuts or gouges, abrasions, and center cap condition _____

   Tire damage, tread, and sidewall _____

   Tire wear _____

   Valve stem damage _____

   Incorrect inflation _____

   Lug nut condition _____

4. Remove wheels and relocate to specified location. Note any issues of excessive rust or corrosion between the wheel and the hub.

5. Recommended tire pressure specifications: F_____   R _____

   Set tire pressure to manufacturer specifications.

   Instructor's check _____

6. Reinstall wheels and torque lug nuts to specification.

       Torque specification _____

       Torque pattern (draw)

7. Does this vehicle require a TPM reset or relearn when performing a tire rotation?

      Yes _____   No _____

8. If at TPM reset or relearn is required, determine what steps need to be performed and summarize them.

_____

_____

_____

9. Remove any tools and equipment and clean up the work area.

                                          Instructor's check _____

Lab Worksheet 5-8

Name _____ Station _____ Date _____

Reinstall and Torque Wheel

ASE Education Foundation Correlation

This lab worksheet addresses the following **MLR** task:

4.D.2. Rotate tires according to manufacturer's recommendations including vehicles equipped with tire pressure monitoring systems (TPMS). **(P-1)**

Description of Vehicle

Year _____ Make _____ Model _____

Engine _____ AT MT CVT PSD (circle which applies)

Procedure

1. Locate and record the wheel lug nut torque specs. _____

2. Draw the recommended pattern in which the lug nuts are tightened.

3. Inspect the wheel studs and hub before setting the wheel in place. Note any signs of damage to the studs or corrosion on the hub. _____

   _____

4. Inspect the wheel lug bore and centerbore for damage. Note your findings. _____

   _____

5. Inspect the lug nuts for internal damage to the thread (external threads if applicable for certain vehicles) and for external damage to the nut. Note your findings _____

   _____

6. If all studs and lug nuts are undamaged and usable, place the wheel onto the hub and studs. Start each lug nut by hand or using a socket by hand only. Seat each lug nut so that the wheel centers and seats properly on the hub.

   Instructor's check _____

7. Set the torque wrench to the proper setting for the lugs.

   Instructor's check _____

8. Tighten each lug in sequence and repeat to ensure all lugs are fully seated and tight.

Instructor's check _____

9. Reinstall hub caps or center caps as necessary.

10. Remove any tools and equipment and clean up the work area.

Instructor's check _____

Lab Worksheet 5-9

Name _____ Station _____ Date _____

Dismount Tire

ASE Education Foundation Correlation

This lab worksheet addresses the following **MLR** task:

4.D.3. Dismount, inspect, and remount tire on wheel; balance wheel and tire assembly. **(P-1)**

Description of Vehicle

Year _____ Make _____ Model _____

Engine _____ AT MT CVT PSD (circle which applies)

Procedure

1. Inspect the wheel and tire for any visible damage and note your findings. _____

   _____

2. Deflate the tire. This can be done by removing the tire valve core or by using a deflation tool.

   Instructor's check _____

3. Break the beads loose and mount the wheel and tire on the tire machine.

   Instructor's check _____

4. Remove the tire from the wheel.

   Instructor's check _____

5. Inspect the tire for evidence of internal damage. Note your findings. _____

   _____

6. Inspect the bead, sidewall, and tread for punctures, excessive wear, or damage. Note your findings.

   _____

   _____

7. Inspect the wheel assembly for excessive rust or corrosion, cracks, and other damage. Note your findings.

   _____

   _____

8. Lubricate the beads and mount the tire back on the wheel.

Instructor's check _____

9. Determine the maximum inflation pressure of the tire.   Maximum pressure _____

10. Inflate the tire and seat the beads. Do not allow the pressure to exceed the maximum inflation pressure.

Instructor's check _____

11. Explain why it is dangerous for the tire to be over-inflated._____

_____

_____

12. Reinstall the valve core (if removed) and inflate to the tire pressure specified on the vehicle.
    Recommended pressure _____

13. Remove any tools and equipment and clean up the work area.

Instructor's check _____

Lab Worksheet 5-10

Name _____ Station _____ Date _____

Balance Wheel and Tire

ASE Education Foundation Correlation

This lab worksheet addresses the following **MLR** task:

4.D.3. Dismount, inspect, and remount tire on wheel; balance wheel and tire assembly. **(P-1)**

Description of Vehicle

Year _____ Make _____ Model _____

Engine _____ AT MT CVT PSD (circle which applies)

Procedure

1. Inspect the wheel and tire for any visible damage. Note your findings. _____
   _____

2. What types of damage to the wheel and tire will affect the wheel and tire balance? _____
   _____

3. Remove the center cap if necessary and mount the wheel and tire to the balancer. Spin the tire by hand and note any excessive lateral (side-to-side) or radial (up and down) movement. Excessive lateral or radial movement may indicate a problem with the wheel and tire or may indicate improper mounting on the balancer. Double check that the wheel and tire is correctly mounted to the balancer before continuing.

4. Input the wheel and tire information into the balancer as necessary.

   Wheel diameter _____

   Wheel width _____

   Wheel offset _____

5. Start the balancer and note the required weight correction amounts.

   Inside weight _____   Outside weight _____

   If the wheel and tire is out of balance, remove any wheel weights and recheck.

   Inside weight _____   Outside weight _____

6. Install the correct weight and recheck tire balance.   OK _____   Not OK _____

Instructor's check _____

7. If the wheel and tire is still out of balance, what actions are necessary to bring it into balance? _____
_____
_____

8. Remove any tools and equipment and clean up the work area.

Instructor's check _____

Lab Worksheet 5-11

Name _____ Station _____ Date _____

Dismount Tire with TPMS

ASE Education Foundation Correlation

This lab worksheet addresses the following **MLR** tasks:

4.D.4. Dismount, inspect, and remount tire on wheel equipped with tire pressure monitoring system sensor. **(P-1)**

4.D.8. Demonstrate knowledge of steps required to remove and replace sensors in a tire pressure monitoring system (TPMS) including relearn procedure. **(P-1)**

Description of Vehicle

Year _____ Make _____ Model _____

Engine _____ AT MT CVT PSD (circle which applies)

TPMS sensor type (circle which applies):   Metal stem   Rubber stem   Strapped

Procedure

1. Inspect the wheel and tire for any visible damage. Note your findings. _____
_____

2. Deflate the tire. This can be done by removing the tire valve core or by using a deflation tool.

    Instructor's check _____

3. Determine how to break the beads loose without damaging the TPM sensor. Break the beads loose and mount the wheel and tire on the tire machine.

    Instructor's check _____

4. Remove the tire from the wheel.

    Instructor's check _____

5. Inspect the tire for evidence of internal damage. Note your findings._____
_____

6. Inspect the bead, sidewall, and tread for punctures, excessive wear, or damage. Note your findings.
_____
_____

7. Inspect the wheel assembly for excessive rust or corrosion, cracks, and other damage. Note your findings.

_____

_____

8. Inspect the TPM sensor for damage. Note your findings. _____

9. Describe how to replace the TPM sensor.

_____

10. Lubricate the bead areas of the tire and mount the tire back on the wheel.

Instructor's check _____

11. Determine the maximum inflation pressure of the tire.   Maximum pressure _____

12. Inflate the tire and seat the beads. Do not allow the pressure to exceed the maximum inflation pressure.

Instructor's check _____

13. Explain why it is dangerous for the tire to be over-inflated. _____

_____

_____

14. Reinstall the valve core (if removed) and set the tire pressure according to the tire pressure decal on the vehicle onto which the tire will be installed.

Recommended pressure _____

15. Verify proper operation of the TPM sensor.

Instructor's check _____

16. Remove any tools and equipment and clean up the work area.

Instructor's check _____

Lab Worksheet 5-12

Name _____ Station _____ Date _____

Inspect Tire for Air Loss

ASE Education Foundation Correlation

This lab worksheet addresses the following **MLR** task:

4.D.5. Inspect tire and wheel assembly for air loss; determine necessary action. **(P-1)**

Description of Vehicle

Year _____ Make _____ Model _____

Engine _____ AT MT CVT PSD (circle which applies)

Tire size _____ Wheel type _____

Procedure

1. Check and record tire pressure. _____

   Recommended tire pressure _____

   If necessary, inflate tire to proper pressure before continuing.

2. Carefully inspect the tire tread section, sidewall, and wheel for signs of damage. Note your findings.

   _____
   _____

3. If your visual inspection does not reveal an obvious cause for air loss, use either a tire dip tank, or a spray bottle with a soapy water solution to locate the leak. Note your findings. _____

   _____

4. If the tire has a puncture in the tread and it is within the repairable area, what type of repair should be made?

   _____
   _____

5. If the tire is leaking around the bead area, what actions are necessary to attempt to repair the wheel and tire?

   _____
   _____

6. If the leak is from or around the valve stem, what steps are required to repair the leak? _____

_____

7. If the tire is leaking from sidewall damage or dry rot, what is required? _____

_____

8. Remove any tools and equipment and clean up the work area.

Instructor's check _____

Lab Worksheet 5-13

Name _____ Station _____ Date _____

Tire Repair

ASE Education Foundation Correlation

This lab worksheet addresses the following **MLR** task:

4.D.6. Repair tire following vehicle manufacturer approved procedure. **(P-1)**

Description of Vehicle

Year _____ Make _____ Model _____

Engine _____ AT MT CVT PSD (circle which applies)

Procedure

1. Using the service information, determine if the tire is repairable. Note your findings.

   _____

   _____

2. What types of tire damage are not repairable? _____

   _____

   _____

3. In what part of the tire can punctures be repaired? _____

   _____

   _____

4. Using a soapy water solution or a tire dip tank, locate the puncture. Use a grease pencil to mark its location.
   Puncture location _____

5. Mark the tire location to the valve stem. Remove any wheel weights. Dismount the tire from the wheel.

   Instructor's check _____

6. Depending on the size of the puncture, you may need to use a reaming tool to open up the hole and allow the stem plug and patch to be installed.

   Instructor's check _____

7. Mount the tire in a tire service clamp and spread the sidewalls apart to gain access to the undertread section of the tire. Use a buffing tool to buff out any texture on the liner around the puncture. The buffed area should be slightly larger than the size of the patch.

Instructor's check _____

8. Clean the buffed area to remove any rubber dust. The patch needs to have a clean surface to adhere properly.

Instructor's check _____

9. Apply a light coat of rubber cement to the buffed area of the liner where the patch will be placed. Refer to the directions on the rubber cement to determine how long the cement needs to dry before applying the patch.

Drying time _____

Instructor's check _____

10. Once the rubber cement has dried, peal the protective film from the patch. Insert the plug through the puncture and pull the patch into place against the liner. Use a stitching tool to work the patch against the liner and ensure no air is trapped between the patch and the tire.

Instructor's check _____

11. Reinstall the tire onto the wheel and air it to the recommended air pressure.

Instructor's check _____

12. Using soapy water, check for leaks around the plug sticking out of the tread. If no leak is found, balance the tire and reinstall on the vehicle.

Instructor's check _____

13. Once finished, put away all tools and equipment and clean up the work area.

Instructor's check _____

Lab Worksheet 5-14

Name _____ Station _____ Date _____

TPMS Service

ASE Education Foundation Correlation

This lab worksheet addresses the following **MLR** task:

4.D.7. Identify indirect and direct tire pressure monitoring systems (TPMS); calibrate system; and verify operation of instrument panel lamps. **(P-1)**

Description of Vehicle

Year _____ Make _____ Model _____

Engine _____ AT MT CVT PSD (circle which applies)

TPMS type _____

Procedure

1. Perform TPM system checks.

   a. TPMS light/indicator operates KOEO.   OK _____ Not OK _____

   b. Start engine and note TPMS light/indicator.   On _____   Off _____   Blinking _____

   c. What does the light indicate? _____

   d. Is actual tire pressure displayed?   Yes _____   No _____

      If Yes, record displayed readings.   RF_____ LF _____ RR _____ LR

   e. _____

2. For direct systems: follow the tire pressure sensor activation sequence using the TPM tool. Record the sensor data.

   LF pressure, temperature, and ID _____

   RF pressure, temperature, and ID _____

   RR pressure, temperature, and ID _____

   LR pressure, temperature, and ID _____

   Spare pressure, temperature, and ID _____

3. Compare the TPMS readings to a tire pressure gauge.

Gauge readings: LF _____   RF _____   LR _____   RR _____

Are the sensor and gauge readings very close to each other?   Yes _____   No _____

What could cause the readings to be different? _____

_____

4. Does the TPMS require a relearn procedure after a tire rotation?   Yes _____   No _____

If yes, describe the relearn procedure. _____

_____

_____

5. Remove any tools and equipment and clean up the work area.

Instructor's check _____

Lab Worksheet 5-15

Name _____ Station _____ Date _____

Reinstall and Torque Wheel

ASE Education Foundation Correlation

This lab worksheet addresses the following **MLR** task:

5.A.3. Install wheel and torque lug nuts. **(P-1)**

Description of Vehicle

Year _____ Make _____ Model _____

Engine _____ AT MT CVT PSD (circle which applies)

Procedure

1. Locate and record the wheel lug nut torque specifications. _____

   _____

2. Draw the recommended pattern in which the lug nuts are tightened.

3. Inspect the wheel studs and hub before setting the wheel in place. Note any signs of damage to the studs or corrosion on the hub. _____

   _____

4. Inspect the wheel lug bore and centerbore for damage. Note your findings. _____

   _____

5. Inspect the lug nuts for internal damage to the thread (external threads if applicable for certain vehicles) and for external damage to the nut. Note your findings. _____

   _____

6. If all studs and lug nuts are undamaged and usable, place the wheel onto the hub and studs. Start each lug nut by hand or using a socket by hand only. Seat each lug nut so that the wheel centers and seats properly on the hub.

   Instructor's check _____

7. Set the torque wrench to the proper setting for the lugs.

Instructor's check _____

8. Tighten each lug in sequence and repeat to ensure all lugs are fully seated and tight.

Instructor's check _____

9. Reinstall hub caps or center caps as necessary.

10. Remove any tools and equipment and clean up the work area.

Instructor's check _____

Lab Worksheet 5-16

Name _____ Station _____ Date _____

Service Wheel Bearing

ASE Education Foundation Correlation

This lab worksheet addresses the following **MLR** tasks:

5.F.1. Remove, clean, inspect, repack, and install wheel bearings; replace seals; and install hub and adjust bearings. **(P-1)**

5.F.5. Replace wheel bearing and race. **(P-2)**

Description of Vehicle

Year _____ Make _____ Model _____

Engine _____ Front bearing _____ Rear bearing _____

Procedure

Safely raise and support the vehicle.

1. Refer to the manufacturer's service information to service the wheel bearings on this vehicle. Summarize the procedures. _____

   _____

   _____

2. Remove the brake caliper and pads. Use a wire hanger or similar device to support the caliper. *Do not allow the caliper to hang by the brake hose.* Remove the bearing dust cap, cotter pin, retainer, washer, and outer wheel bearing from the hub. Describe the condition of the grease and the outer bearing components.____

   _____

   _____

3. Remove the brake rotor, inner wheel bearing, and grease seal. Describe how you removed the inner bearing and seal. _____

   _____

4. Clean the hub, bearings, and bearing components following the manufacturer's service procedures. Once the parts are clean and dry, inspect the parts for wear and damage. Note your findings. _____

   _____

   _____

5. To remove a bearing race, use a brass punch and hammer to drive the race from the hub. To install a new race, use a brass or aluminum bearing driver to prevent damage to the race.

Instructor's check _____

6. Determine the correct grease to repack the bearings.

Bearing grease specification _____

7. Pack the bearings with grease until no air pockets remain. Apply a light coating of grease to the spindle, bearing races, and hub.

Instructor's check _____

8. Reinstall the inner wheel bearing into the race and place a new grease seal over the bearings. Use a seal driver to fully seat the seal into the hub.

Instructor's check _____

9. Install the rotor onto the spindle and install the outer wheel bearing. Install the washer and hand tighten the spindle nut.

Instructor's check _____

10. Tighten and adjust the wheel bearing as described in the service information. Record the procedure to correctly adjust the wheel bearing. _____

_____

_____

Instructor's check _____

11. What can result from the wheel bearing not being adjusted properly? _____

_____

_____

12. Install the cotter pin and secure it to prevent the spindle nut from working loose.

Instructor's check _____

13. Place a small amount of grease in the bearing cap and install the cap.

Instructor's check _____

14. Reinstall the caliper and brake pads. Torque all fasteners to specifications.

Instructor's check _____

15. Once finished, put away all tools and equipment and clean up the work area.

Instructor's check _____

Lab Worksheet 5-17

Name _____ Station _____ Date _____

Inspect and Replace Wheel Studs

ASE Education Foundation Correlation

This lab worksheet addresses the following **MLR** task:

5.F.6. Inspect and replace wheel studs. **(P-1)**

Description of Vehicle

Year _____ Make _____ Model _____

Engine _____ AT MT CVT PSD (circle which applies)

Procedure

Safely raise and support the vehicle.

1. Begin your inspection by checking all lug nuts and studs for signs of damage, for missing lug nuts and studs, and if all lug nuts and studs are of the same type and size. Summarize your findings. _____

    _____

    _____

2. Select the appropriate size socket(s) and loosen all of the lug nuts and studs. Note if any of the lug nuts and studs require more or less effort than others as this can indicate improper torque and damage to the lug nuts or studs. Note your findings. _____

    _____

    _____

3. Remove the lug nuts and studs and set them aside for inspection. Remove each wheel and tire assembly. Inspect all of the lug nuts and studs for signs of damage. Note your findings. _____

    _____

    _____

    If the threads are badly damaged or the stud is broken, you will have to replace the stud. Refer to the manufacturer's service information for the correct procedure as it may require removing the hub or other components to access the studs.

4. Remove the damaged wheel stud from the axle hub.

Instructor's check _____

5. Compare the replacement stud to the original. Ensure it is the correct length, diameter, and thread pitch.

6. Insert the stud into the axle hub. There are several methods to fully seat the wheel stud. Describe the tools and method used. _____

_____

7. Thread a lug nut onto the new stud and verify proper operation.

Instructor's check _____

8. Once finished, put away all tools and equipment and clean up the work area.

Instructor's check _____

CHAPTER 6

Suspension System Principles

Review Questions

1. All of the suspension components must work together to provide the _____ quality and _____ characteristics expected by the driver and passengers.

2. Summarize six functions of the suspension system.

   a. _____

   b. _____

   c. _____

   d. _____

   e. _____

   f. _____

3. The _____ of the vehicle is carried through the springs to the wheels and tires.

4. Vehicles with a body-over-frame design often have a _____ shaped frame.

5. _____ are frame components that attach to the frame rails and carry the engine and transmission.

6. Rubber _____ are used to isolate the frame from the body.

7. Most modern vehicles use a _____ type frame.

8. A(n) _____ suspension allows each wheel to move separately for the best ride quality.

9. The type of front suspension used on many heavy-duty vehicles is the:

   a. Independent

   b. I-beam

   c. 4WD

   d. None of the above

10. A dependent rear axle that drives the wheels is called a _____ axle.

11. Used on the rear of many FWD cars, a _____ axle is a type of dependent rear suspension.

12. A semi-independent rear axle is able to _____ slightly, providing improved ride and handling.

13. During braking, as much as _____% of the weight is transferred to the front of the vehicle.

14. If the rear tires of a vehicle reach their grip and cornering limits before the front tires, it is called _____.

15. When the front of the vehicle cannot make a turn through the desired turn radius because the front tires have lost traction, it is called _____.

16. When a coil spring _____, it stores energy.

17. Coil springs often use rubber _____ between the spring and the frame to reduce noise.

18. Describe the difference between how standard and variable-rate coil springs operate.

    _____

    _____

19. Leaf springs are attached to the frame with a _____ or bracket.

20. Adding additional leaf springs allows the vehicle to carry _____ weight.

21. *Technician A* says air springs are used only on large commercial trucks because they make the vehicle ride very rough. *Technician B* says air springs are adjustable, meaning ride height or load-carrying capacity can vary depending on the air pressure in the springs. Who is correct?
    a. Technician A
    b. Technician B
    c. Both A and B
    d. Neither A nor B

22. Torsion bars _____ when the control arm moves in response to road conditions.

23. Torsion bars are most commonly used on what type of vehicle?
    a. FWD
    b. AWD
    c. 4WD
    d. RWD

24. The upward movement of the tire is called _____ and the downward movement of the tire is called _____.

25. *Technician A* says that the weight carried by the springs is called sprung weight. *Technician B* says sprung weight includes the wheels and tires. Who is correct?
    a. Technician A
    b. Technician B
    c. Both A and B
    d. Neither A nor B

26. Shock absorbers are used to _____ the spring oscillations.

27. As the shock piston moves up and down in the main chamber, oil is _____ into a second chamber.

28. Control arms are also called _____ arms and _____ due to their similarity to those shapes.

29. Ball joints allow the steering knuckle to _____ while providing a tight connection to the control arms.

30. Describe the two types of ball joints. _____
    _____
    _____

31. List four methods by which ball joints are attached to control arms. _____
    _____

32. Describe the purpose and operation of the stabilizer bar. _____
    _____
    _____

33. Explain the construction of a MacPherson strut. _____
    _____
    _____

34. A _____ strut locates the spring separate from the shock.

35. Explain how the strut differs between a multilink and MacPherson strut suspension.
    _____
    _____

36. List the components commonly used in a short/long arm suspension system.

_____

_____

37. A 4WD suspension may use a solid _____ axle or it may mount the differential to the chassis to allow for an _____ front suspension.

38. List at least five components that are typically unsprung weight. _____

_____

_____

39. What components are eliminated by the MacPherson strut suspension?

_____

40. *Technician A* says the sprung weight of the vehicle is carried by the lower ball joint on a MacPherson strut suspension. *Technician B* says the sprung weight of the vehicle is carried by the lower ball joint on a modified strut suspension. Who is correct?

    a. Technician A

    b. Technician B

    c. Both A and B

    d. Neither A nor B

41. Many 4WD trucks and SUVs use _____ _____ instead of coil springs.

42. Wind-up occurs when _____ is applied to the rear axle.

43. Many rear suspension systems use a _____ bar or a _____ bar to limit rear axle movement.

44. Magneride uses a _____ fluid that can change viscosity.

Activities

1. Identify the following suspension types.

Figure 6-1   Figure 6-2   Figure 6-3   Figure 6-4   Figure 6-5

SLA   MacPherson strut   Modified strut   Multilink   Live axle

       Semi-independent axle   Torsion bar   Independent4WD

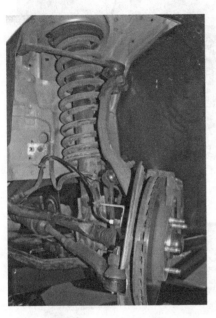

Figure 6-1

Figure 6-2

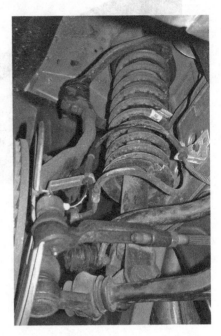

Figure 6-3

Figure 6-4

Figure 6-5

2. Identify the types of springs shown below.

Figure 6-6	Figure 6-7	Figure 6-8	Figure 6-9	Figure 6-10
Variable-rate coil	Standard-rate coil	Leaf	Torsion bar	Air

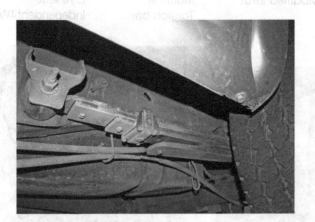

Figure 6-6

Figure 6-7

Figure 6-8

Figure 6-9

Figure 6-10

3. Circle the components that are the unsprung weight in Figure 6-11.

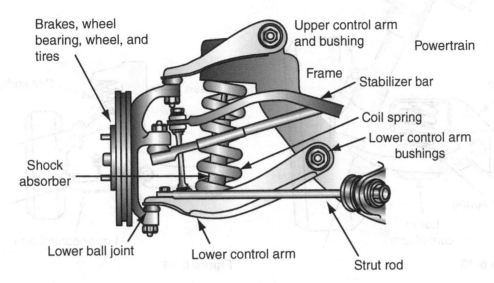

Brakes, wheel bearing, wheel, and tires

Upper control arm and bushing

Powertrain

Frame

Stabilizer bar

Coil spring

Lower control arm bushings

Shock absorber

Lower ball joint

Lower control arm

Strut rod

Figure 6-11

4. Label the load-carrying ball joint in the following figures.

Figure 6-12

SLA suspension

Figure 6-13

Modified strut suspension

Figure 6-14

Multilink suspension

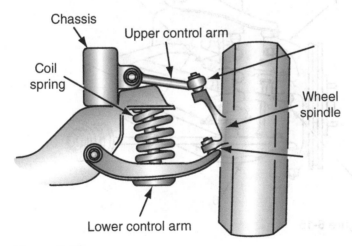

Chassis

Upper control arm

Coil spring

Wheel spindle

Lower control arm

Figure 6-12

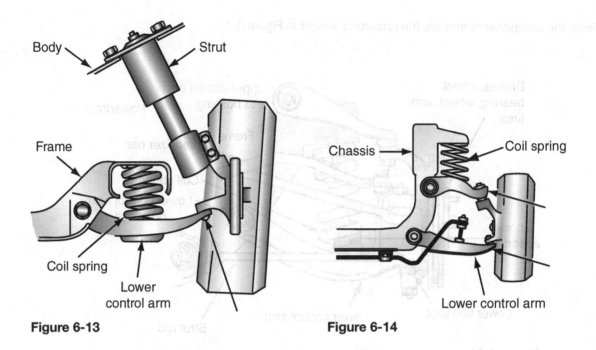

Body

Strut

Frame

Chassis

Coil spring

Coil spring

Lower
control arm

Lower control arm

Figure 6-13

Figure 6-14

5. Label the components of the suspension shown below.

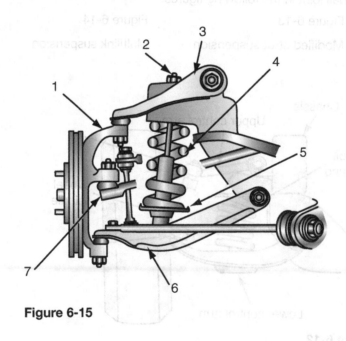

Figure 6-15

6. Match the component to the descriptions.

Spring	Connects the control arm to the stabilizer bar
Steering knuckle	Dampens spring oscillations
Shock absorber	Carries weight of vehicle
Upper control arm	Combines shock and spring
Lower control arm	Supports spring and load-carrying ball joint
Control arm bushing	Connects upper and lower control arms
Strut	Reduces body roll
Bearing plate	Typically carries following ball joint
Cam bolt	Allows strut to pivot
Strut rod	Used to adjust camber
Sway bar	Sometimes used to adjust caster
Sway bar links	Allows for control arm movement

Lab Activity 6-1

Name _____ Date _____ Instructor _____

Year _____ Make _____ Model _____

Identify suspension types and components—short/long arm

1. Locate the control arms and describe the shape of each.
   Upper control arm _____
   Lower control arm _____

2. How many ball joints are used? _____

3. What is the location of the load-carrying ball joint? _____

4. How are the ball joints secured to the control arms?
   Upper ball joint _____
   Lower ball joint _____

5. What types of springs are used and where are they located? _____
   _____

6. Where are the shock absorbers mounted and how? _____
   _____

7. Describe the type of stabilizer bar links used. _____

8. How are changes to caster and camber made on this suspension? _____

9. Does the suspension use electronic ride control?   _____ Yes   _____ No
   If yes, describe the system and its basic operation. _____
   _____
   _____

10. Using service information, locate a TSB regarding the front suspension of this vehicle or of a vehicle with a similar suspension system. Record the TSB number, title, and a brief summary of the subject of the TSB.
    _____
    _____
    _____

Lab Activity 6-2

Name _____ Date _____ Instructor _____

Year _____ Make _____ Model _____

Identify suspension types and components—MacPherson strut

1. Locate the control arms and describe the shape of each.

    Lower control arm _____

2. How many ball joints are used? _____

3. How is the strut connected to the steering knuckle? _____

4. How are the ball joints secured to the control arms?

    Lower ball joint _____

5. What types of springs are used and where are they located? _____

    _____

6. Where are the shock absorbers mounted and how? _____

    _____

7. Describe the type of stabilizer bar links used. _____

8. How are changes to caster and camber made on this suspension? _____

    _____

9. Does the suspension use electronic ride control?   _____ Yes   _____ No

    If yes, describe the system and its basic operation. _____

    _____

    _____

10. Using service information, locate a TSB regarding the front suspension of this vehicle or of a vehicle with a similar suspension system. Record the TSB number, title, and a brief summary of the subject of the TSB.

    _____

    _____

    _____

Lab Activity 6-3

Name _____ Date _____ Instructor _____

Year _____ Make _____ Model _____

Identify suspension types and components—multilink

1. Locate the control arms and describe the shape of each.
   Upper control arm _____
   Lower control arm _____

2. How many ball joints are used? _____

3. What is the location of the load-carrying ball joint? _____

4. How are the ball joints secured to the control arms?
   Upper ball joint _____
   Lower ball joint _____

5. What types of springs are used and where are they located? _____
   _____

6. Where are the shock absorbers mounted and how? _____
   _____

7. Describe the type of stabilizer bar links used. _____

8. How are changes to caster and camber made on this suspension? _____

9. Describe how the steering knuckle used in this suspension is different from steering knuckles used in MacPherson strut and SLA suspensions. _____
   _____

10. Does the suspension use electronic ride control? _____ Yes   _____ No
    If yes, describe the system and its basic operation. _____
    _____
    _____

11. Using service information, locate a TSB regarding the front suspension of this vehicle or of a vehicle with a similar suspension system. Record the TSB number, title, and a brief summary of the subject of the TSB.

_____

_____

_____

Lab Activity 6-4

Name _____ Date _____ Instructor _____

Year _____ Make _____ Model _____

Identify suspension types and components—4WD

1. Describe the type of front suspension used on this vehicle. _____
   _____
   _____

2. Does this suspension use control arms?   _____ Yes   _____ No

3. How many ball joints are used? _____

4. What is the location of the load-carrying ball joint, if applicable? _____
   _____

5. How are the ball joints secured to the control arms?
   Upper ball joint _____
   Lower ball joint _____

6. What types of springs are used and where are they located? _____
   _____

7. Where are the shock absorbers mounted and how? _____
   _____

8. Describe the type of stabilizer bar links used. _____
   _____

9. How are changes to caster and camber made on this suspension? _____

10. Using service information, locate a TSB regarding the front suspension of this vehicle or of a vehicle with a similar suspension system. Record the TSB number, title, and a brief summary of the subject of the TSB.
    _____
    _____
    _____

Lab Activity 6-5

Name _____ Date _____ Instructor _____

Year _____ Make _____ Model _____

Identify suspension types and components—rear suspension

1. What type of rear suspension system is used on this vehicle?_____
   Check all components that apply:
   Upper control arm _____
   Lower control arm _____
   Strut _____

2. How many ball joints are used? _____

3. What is the location of the load-carrying ball joint? _____

4. How are the ball joints secured to the control arms?
   Upper ball joint _____
   Lower ball joint _____

5. What types of springs are used and where are they located? _____
   _____

6. Where are the shock absorbers mounted and how? _____
   _____

7. Describe the type of stabilizer bar links used. _____
   _____

8. How are changes to camber made on this suspension? _____

9. Does the suspension use electronic ride control? _____ Yes _____ No
   If yes, describe the system and its basic operation. _____
   _____
   _____

10. Using service information, locate a TSB regarding the rear suspension of this vehicle or of a vehicle with a similar suspension system. Record the TSB number, title, and a brief summary of the subject of the TSB.
    _____
    _____
    _____

CHAPTER 7

Suspension System Service

Review Questions

1. Describe five tools commonly used in suspension system service and repair.

   a. _____

   b. _____

   c. _____

   d. _____

   e. _____

2. Always use your _____ muscles when lifting.

3. Working on suspension components often means working with _____ components and fasteners.

4. Before beginning any work, make sure the vehicle is properly raised and _____.

5. Before using a strut compressor, make sure you are fully _____ in its use and understand how to safely operate the equipment.

6. _____ should never be applied to a shock absorber since it can cause the shock to rupture.

7. Inspection of the suspension system also includes looking at the _____ and the _____ system.

8. In every aspect of automotive service and repair, you first should _____ the customer's complaint.

9. Before any repairs can be made, a thorough _____ of the suspension system must be performed.

10. As the springs age and weaken, vehicle _____ decreases.

11. Describe how vehicle ride and handling are affected by weak springs.

_____

_____

12. Explain how to check vehicle ride height. _____

_____

13. *Technician A* says weak shocks will affect ride height. *Technician B* says weak shocks will affect ride quality. Who is correct?

    a. Technician A

    b. Technician B

    c. Both A and B

    d. Neither A nor B

14. *Technician A* says installing smaller wheels and tires will affect ride height. Technician B says low tire pressure will affect ride height. Who is correct?

    a. Technician A

    b. Technician B

    c. Both A and B

    d. Neither A nor B

15. Which of the following is not a symptom of a worn-out shock?

    a. Noise when bouncing

    b. Oil leakage

    c. Excessive bouncing

    d. Incorrect ride height

16. True or False: A vehicle may show more than one type of tire wear. _____

17. A tire shows excessive wear in the center of the tread. Which is the most likely cause?

    a. Underinflation

    b. Overinflation

    c. Negative camber

    d. Positive camber

18. Noises can be caused by worn suspension components, such as loose _____ _____ and strut _____ plates.

19. A _____ is often necessary when trying to determine the cause of a noise or vibration.

20. Which of the following components has the primary purpose of reducing body sway?
    a. Shock absorber
    b. Control arm bushing
    c. Stabilizer bar
    d. Rebound bumper

21. You should also perform a search of the vehicle's service _____ and for any relevant technical service _____.

22. The most common suspension type in use today is the _____ suspension.

23. Label the components of the MacPherson strut shown in Figure 7-1.

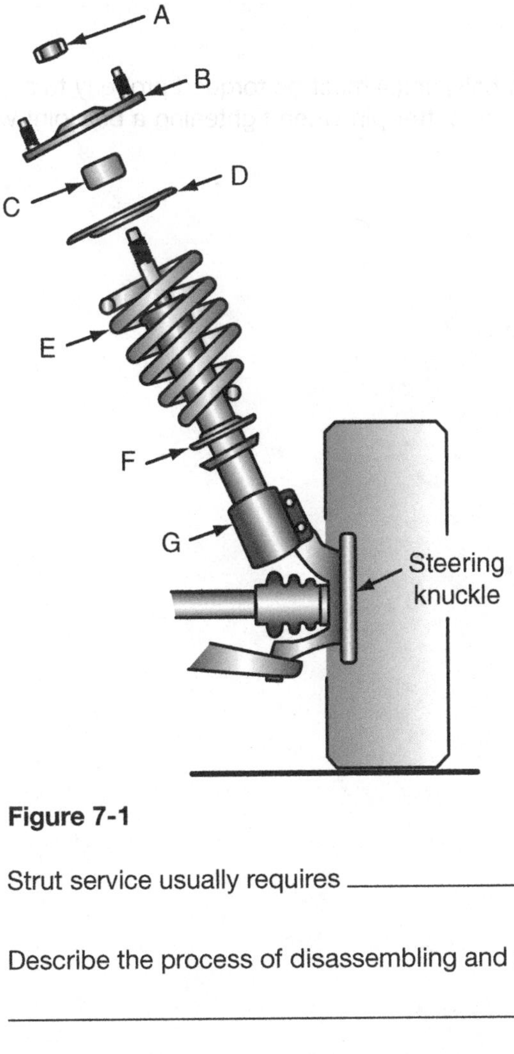

a. _____

b. _____

c. _____

d. _____

e. _____

f. _____

g. _____

Figure 7-1

24. Strut service usually requires _____ the strut from the vehicle.

25. Describe the process of disassembling and reassembling a MacPherson strut.

_____

_____

_____

_____

26. In some vehicles, the strut _____ can be replaced while the strut is still installed in the vehicle.

27. When you are replacing gas-charged struts or shocks, the gas charge needs to be _____ before throwing the strut or shock away.

28. When replacing shock absorbers, sometimes a _____ is needed to support the axle.

29. A bumper that appears _____ means that it has been contacting the frame, which can indicate a weak _____.

30. Which of the following is often necessary when replacing a steering knuckle?
    a. Removing the shock absorber
    b. Separating a ball joint
    c. Removing the tie rod
    d. All of the above

31. When replacing a steering knuckle: *Technician A* says the ball joint(s) must be torqued properly to support the vehicle's weight. *Technician B* says always install a new cotter pin when tightening a ball joint with a castle nut. Who is correct?
    a. Technician A
    b. Technician B
    c. Both A and B
    d. Neither A nor B

32. What is the purpose of the tool shown in Figure 7-2?

Figure 7-2

    a. Removing wheel studs
    b. Installing ball joints
    c. Removing an axle
    d. Separating ball joints

33. Label the jacking and supporting locations to check ball joint wear for the suspensions shown in Figure 7-3.

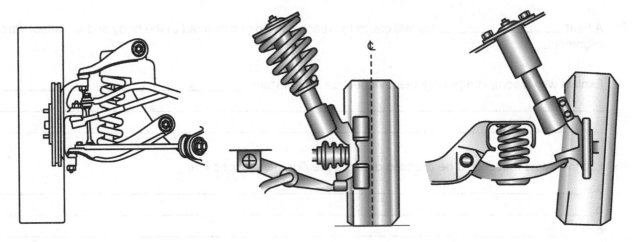

Figure 7-3

34. Explain the typical procedure to check ball joint wear for a load-carrying ball joint mounted in a lower control arm. _____

_____

_____

35. Some ball joints have built-in _____ that change how far the grease fitting protrudes from the base of the joint.

36. To remove and install ball joints on many types of vehicles, a ball joint _____ is necessary.

37. On some vehicles, it is necessary to replace the _____ _____ in order to replace the ball joint.

38. Control arm _____ replacement often requires special bushing tools.

39. Describe the problems caused by loose and worn radius arm bushings. _____

_____

_____

40. Explain the purpose of the sway bar in the front suspension. _____

_____

41. Grease or _____ fittings are used to lubricate suspension and steering joints.

42. Many rear struts bolt either into the _____ or into the rear firewall area where the rear seat-back and parcel shelf are located.

43. When you are replacing the rear _____ the rear suspension may need to be supported.

44. A bent _____ bar will cause the rear axle to be misaligned to the body and will affect wheel alignment.

45. Explain why a complete prealignment inspection is important. _____
    _____
    _____

46. List five items that should be checked during a prealignment inspection.
    a. _____
    b. _____
    c. _____
    d. _____
    e. _____

47. Describe how and why to perform a dry-park check. _____
    _____
    _____
    _____

48. The imaginary line that bisects the vehicle is called the _____
    _____.

49. The _____ angle is tilt of the tire away from vertical.

50. The _____ angle is responsible for tracking and steering return.

51. Positive toe or toe _____ is when the front of the tires are closer together than the distance at the rear of the tires.

52. List the proper sequence of alignment adjustments in order. _____
    _____
    _____

Activities

1. Label the tools shown in Figure 7-4 (a through f).

a.

b.

c.

d.

e.

f.

Figure 7-4

2. For the suspension shown in Figure 7-5, identify the components which, if excessively worn, will adversely affect wheel alignment.

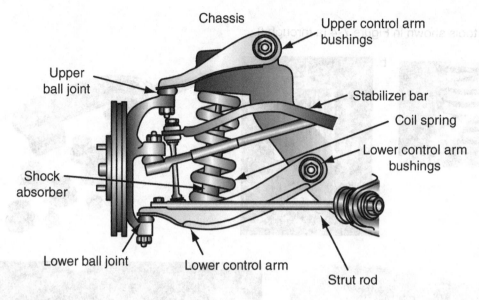

Chassis

Upper control arm bushings

Upper ball joint

Stabilizer bar

Coil spring

Lower control arm bushings

Shock absorber

Lower ball joint

Lower control arm

Strut rod

Figure 7-5

_____

_____

3. Match the following complaints with their most likely causes.

Excessive body lean	Worn shocks
Creaking noise	Worn control arm bushing
Knocking over bumps	Worn ball joint
Excessive bounce	Broken sway bar link
Low ride height	Weak spring

Lab Activity 7-1

PREALIGNMENT INSPECTION CHECKLIST

Owner _____ Phone _____ Date _____

Address _____ VIN _____

Make _____ Model _____ Year _____ Lic. number _____ Mileage _____

1. Road test results	Yes	No	Right	Left
Above 30 MPH				
Below 30 MPH				
Bump steer				
When bracking				
Steering wheel movement				
Stopping from 2–3 MPH (Front)				
Vehicle steers hard				
Strg wheel returnability normal				
Strg wheel position				
Vibration	Yes	No	Frnt	Rear

2. Tire pressure	Specs Frnt _____ Rear _____

Record pressure found

RF _____ LF _____ RR _____ LR _____

3. Chassis height	Specs Frnt _____ Rear _____

Record height found

RF _____ LF _____ RR _____ LR _____

	Yes	No
Springs sagged		
Torsion bars adjusted		

4. Rubber bushings	OK
Upper control arm	
Lower control arm	
Sway bar/stabilizer link	
Strut rod	
Rear bushing	

5. Shock absorbers/struts	Frnt	Rear

6. Steering linkage	Frnt OK	Rear OK
Tie-rod ends		
Idler arm		
Center link		
Sector shaft		
Pitman arm		
Gearbox/rack adjustment		
Gearbox/rack mounting		

7. Ball joints			OK
Load bearings			
Specs		Readings	
Right _____ Left _____		Right _____ Left _____	
Follower			
Upper strut bearing mount			
Rear			

8. Power steering	OK
Belt tension	
Fluid level	
Leaks/hose fittings	
Spool valve centered	

9. Tires/wheels	OK
Wheel runout	
Condition	
Equal tread depth	
Wheel bearing	

10. Brakes operating properly	

11. Alignment	Spec		Initial reading		Adjusted reading	
	R	L	R	L	R	L
Camber						
Caster						
Toe						

Bump steer	Toe change right wheel		Toe change left wheel	
	Amount	Direction	Amount	Direction
Chassis down 3"				
Chassis up 3"				

	Spec		Initial reading		Adjusted reading	
	R	L	R	L	R	L
Toe-out on turns						
SAI						
Rear camber						
Rear total toe						
Rear indiv. toe						
Wheel balance						
Radial tire pull						

Lab Activity 7-1

PREALIGNMENT INSPECTION CHECKLIST

Owner _____ Phone _____ Date _____

Address _____ VIN _____

Make _____ Model _____ Year _____ Lic. number _____ Mileage _____

1. Road test results	Yes	No	Right	Left
Above 30 MPH				
Below 30 MPH				
Bump steer				
When braking				
Shimmy when unevolved				
Shimmy from 2–3 Mph (Pb mp)				
Vehicle steers hard				
Stg. wheel return ability difficult				
Stg. wheel position				
Vibration				

2. Tire pressure	Speed	Full	Rear
Record pressure found			
RF _____ LF _____ RR _____ LR			

3. Chassis height	Speed	Full	Rear
Record height found			
RF _____ LF _____ RR _____ LR			

	Yes	No
Springs sag /oad		
Torsion bars adjusted		

4. Rubber bushings	OK
Upper control arm	
Lower control arm	
Sway bar/stabilizer link	
Strut rod	
Rear bushing	

5. Shock absorber/struts	LF	RF	LR	RR

6. Steering linkage	RF OK	RR OK	Rear OK
Tie rod ends			
Idler arm			
Center link			
Sector shaft			
Pitman arm			
Gearbox/rack adjustment			
Gearbox/rack mounting			

7. Ball joints						OK
Load bearings						
Shock	Readings					
	Right	Left	Right	Left		
Follower						
Upper/lower bearing (noted)						
Rear						

8. Power steering		OK
Belt tension		
Fluid level		
Leaks/hose fittings		
Spool valve centered		

9. Tires/wheels		OK
Wheel runout		
Condition		
Equal tread depth		
Wheel bearing		

10. Brakes operating properly

11. Alignment	Spec	Initial reading	Adjusted reading
		L	R	L	R	L	R
Camber					
Caster					
Toe					
Bump steer					

	Toe change left wheel	Toe change right wheel
Chassis down 3"		
Chassis up 3"		

	Spec	Initial reading	Adjusted reading
		L	R	L	R	L	R
Toe-out on turns					
SAI					
Rear camber					
Rear total toe					
Rear indiv. toe					
Wheel balance					
Radial tire pull					

Lab Worksheet 7-1

Name _____ Station _____ Date _____

Identify SRS

ASE Education Foundation Correlation

This lab worksheet addresses the following **MLR** task:

4.A.1. Research vehicle service information including fluid type, vehicle service history, service precautions, and technical service bulletins. **(P-1)**

Description of Vehicle

Year _____ Make _____ Model _____

Engine _____ AT MT CVT PSD (circle which applies)

Procedure

1. Determine the type(s) of air bags installed in the vehicle: Check all that apply.

   Driver _____

   Passenger _____

   Side impact_____

   Curtain _____

   Knee _____

   Other _____

2. Does the vehicle have a passenger air bag on/off switch?   Yes _____   No _____

   If yes, describe location _____

3. Does the vehicle have an occupant classification system?   Yes _____ No _____

   If yes, what types of sensors are used? _____

4. Using service information, determine the location of the sensors used by the SRS.

Sensor type _____   Location _____

Sensor type _____   Location _____

Sensor type _____   Location _____

Sensor type _____   Location _____

Lab Worksheet 7-2

Name _____ Station _____ Date _____

Identify Suspension and Steering Systems and Components

ASE Education Foundation Correlation

This lab worksheet addresses the following **MLR** task:

4.A.3. Identify suspension and steering system components and configurations. **(P-1)**

Procedure

1. Identify each of the following types of suspension systems.

   MacPherson strut   Vehicle YMM _____

   Multilink   Vehicle YMM _____

   Short and long arm   Vehicle YMM _____

   4WD live axle   Vehicle YMM _____

   Other   Vehicle YMM _____

2. Identify the suspension system components provided by your instructor.

   Coil spring _____   Torsion bar _____   Leaf spring _____

   Air spring _____   Sway bar _____   Control arm _____

   Ball joint _____   Sway bar link _____   Bushing _____

   Track bar _____   Shock absorber _____   Radius arm _____

3. Identify the steering system components provided by your instructor.

   Steering shaft _____   Coupler/joint _____   Outer tie rod _____

   Inner tie rod _____   Rack and pinion _____   Bellows_____

   Pitman arm _____   Centerlink _____   Idler arm _____

   Power steering pump _____ P/S hoses _____   Drag link _____

   Steering damper _____   EPS motor _____

Lab Worksheet 7-3

Name _____ Station _____ Date _____

Inspect Control Arms

ASE Education Foundation Correlation

This lab worksheet addresses the following **MLR** task:

4.B.9. Inspect upper and lower control arms, bushings, and shafts. **(P-1)**

Description of Vehicle

Year _____ Make _____ Model _____

Engine _____ AT MT CVT PSD (circle which applies)

Procedure

1. Refer to the manufacturer's service information for inspecting suspension system components. Summarize your findings.

   _____

   _____

   _____

2. Inspect the upper and lower control arms for damage such as cracks, rust-through, or being bent. Note your findings.

   _____

   _____

   _____

3. Locate and inspect the control arm bushings. Check for cracks in the rubber and for bushing separation. Note your findings. _____

   _____

   _____

4. With the vehicle on a flat, level surface, have an assistant jounce the front end while you observe the control arms and bushings. Note your findings. _____

   _____

   _____

5. Describe what types of concerns can be caused by loose or damaged control arm components. _____
_____
_____

6. Based on your inspection, what are the necessary actions for this vehicle? _____
_____
_____

7. Once finished, put away all tools and equipment and clean up the work area. _____

Instructor's check _____

Lab Worksheet 7-4

Name _____ Station _____ Date _____

Inspect Rebound Bumpers

ASE Education Foundation Correlation

This lab worksheet addresses the following **MLR** task:

4.B.10. Inspect and replace rebound bumpers. **(P-1)**

Description of Vehicle

Year _____ Make _____ Model _____

Engine _____ AT MT CVT PSD (circle which applies)

Procedure

1. Refer to the manufacturer's service information for inspecting suspension system components. Summarize your findings. _____
   _____
   _____

2. Inspect the upper jounce and rebound bumpers. Note your findings._____
   _____
   _____

3. If the jounce and rebound bumpers show signs of wear or damage, what may be the cause? _____
   _____
   _____

4. With the vehicle on a flat, level surface, have an assistant jounce the vehicle while you observe the movement of the suspension. Note your findings. _____
   _____
   _____

5. Refer to the service information for the steps to replace the jounce or rebound bumper for this vehicle. Summarize the procedure. _____

_____

_____

6. Based on your inspection, what are the necessary actions for this vehicle? _____

_____

_____

7. Once finished, put away all tools and equipment and clean up the work area. _____

Instructor's check _____

Lab Worksheet 7-5

Name _____ Station _____ Date _____

Inspect Rear Suspension Components

ASE Education Foundation Correlation

This lab worksheet addresses the following **MLR** task:

4.B.11. Inspect track bar, strut rods and radius arms, and related mounts and bushings. **(P-1)**

Description of Vehicle

Year _____ Make _____ Model _____

Engine _____ AT MT CVT PSD (circle which applies)

Procedure

1. Refer to the manufacturer's service information for inspecting suspension system components. Summarize your findings.

   _____
   _____
   _____

2. Identify each of the following:

   Track bar_____   Instructor's check _____

   Strut rod _____   Instructor's check _____

   Radius arm _____   Instructor's check _____

3. Inspect the track bar, strut rod, or radius arm for damage such as cracks or being bent. Note your findings.

   _____
   _____
   _____

4. Locate and inspect the track bar, strut rod, or radius arm bushings. Check for cracks in the rubber and for bushing separation. Note your findings. _____

   _____
   _____

5. With the vehicle on a flat, level surface, have an assistant jounce the vehicle while you observe the track bar, strut rod, or radius arms and bushings. Note your findings. _____

_____

_____

6. Describe what types of concerns can be caused by loose or damaged track bars, strut rods, or radius arms and bushings. _____

_____

_____

7. Based on your inspection, what are the necessary actions for this vehicle? _____

_____

_____

Lab Worksheet 7-6

Name _____ Station _____ Date _____

Inspect Ball Joints

ASE Education Foundation Correlation

This lab worksheet addresses the following **MLR** task:

4.B.12. Inspect upper and lower ball joints (with or without wear indicators). **(P-1)**

Description of Vehicle

Year _____ Make _____ Model _____

Engine _____ AT MT CVT PSD (circle which applies)

Procedure

1. Refer to the manufacturer's service information for inspecting the ball joints on this vehicle. Summarize your findings. _____
   _____
   _____

2. For vehicles with wear indicating ball joints, check the protrusion from the joint at the grease fitting. Note your findings.
   _____
   _____
   _____

3. For nonwear indicating ball joints, determine how to support the vehicle to unload the ball joints to check for wear.

   Jack placement _____

                                          Instructor's check _____

4. Many vehicles allow a limited amount of play in the ball joints before they are considered worn. Refer to the service information and determine the allowable play.

   Play specification _____

5. Check for play in the ball joints. Note your findings. _____
   _____
   _____

6. If a certain amount of movement is allowed, you will need to use a dial indicator to measure the play. Set up a dial indicator to measure ball joint play and record your findings.

   Instructor's check _____

   Measured play _____

7. Based on your inspection, what are the necessary actions for this vehicle? _____
   _____
   _____

8. Describe what types of concerns can be caused by loose ball joints. _____
   _____
   _____

9. Based on your inspection, what is the condition of the ball joints? _____
   _____
   _____

10. Once finished, put away all tools and equipment and clean up the work area. _____
    _____
    _____

    Instructor's check _____

Lab Worksheet 7-7

Name _____ Station _____ Date _____

Inspect Coil Springs

ASE Education Foundation Correlation

This lab worksheet addresses the following **MLR** task:

4.B.13. Inspect suspension system coil springs and spring insulators (silencers). **(P-1)**

Description of Vehicle

Year _____ Make _____ Model _____

Engine _____ AT MT CVT PSD (circle which applies)

Procedure

1. Refer to the manufacturer's service information for inspecting the springs on this vehicle. Summarize your findings. _____
   _____
   _____

2. Perform a visual inspection of the springs and insulators. Check for broken springs, worn, damaged, or missing insulators. Note your results._____
   _____
   _____

3. As springs age they often cannot support the weight of the vehicle at the specified ride height. Locate the ride height specifications.

   Ride height specifications _____

4. Measure the ride height as described in the service information. Note your results.
   _____
   _____

5. Based on your inspection, describe the condition of the vehicle and determine what actions are necessary.
   _____
   _____

6. Once finished, put away all tools and equipment and clean up the work area.

   Instructor's check _____

Lab Worksheet 7-8

Name _____ Station _____ Date _____

Inspect Torsion Bars

ASE Education Foundation Correlation

This lab worksheet addresses the following **MLR** task:

4.B.14. Inspect suspension system torsion bars and mounts. **(P-1)**

Description of Vehicle

Year _____ Make _____ Model _____

Engine _____ AT MT CVT PSD (circle which applies)

Procedure

1. Refer to the manufacturer's service information for inspecting the springs on this vehicle. Summarize your findings. _____
_____
_____

2. Perform a visual inspection of the torsion bars and mounts. Check for broken torsion bars, worn, damaged, or missing mounts and insulators. Note your results._____
_____
_____

3. As springs age they often cannot support the weight of the vehicle at the specified ride height. Locate the ride height specifications.

   Ride height specifications _____

4. Measure the ride height as described in the service information. Note your results.
_____
_____

5. Based on your inspection, describe the condition of the vehicle and determine what actions are necessary.
_____
_____
_____

6. Once finished, put away all tools and equipment and clean up the work area.

Instructor's check _____

Lab Worksheet 7-9

Name _____ Station _____ Date _____

Inspect Sway Bars

ASE Education Foundation Correlation

This lab worksheet addresses the following **MLR** task:

4.B.15. Inspect and/or replace front/rear stabilizer bar (sway bar) bushings, brackets, and links. **(P-1)**

Description of Vehicle

Year _____ Make _____ Model _____

Engine _____ AT MT CVT PSD (circle which applies)

Procedure

1. Refer to the manufacturer's service information for inspecting the sway bar, bushings, brackets, and links on this vehicle. Summarize your findings. _____
   _____
   _____

2. Perform a visual inspection of the sway bar components. Check for broken links, worn, damaged, or missing mounts and bushings. Note your results._____
   _____
   _____

3. What customer complaints can be caused by broken, worn, or damaged sway bar components?
   _____
   _____
   _____

4. To replace a broken or damaged sway bar link, first remove the old link. Describe the type of link used on this vehicle and how it is removed._____
   _____
   _____

5. Remove the old link and compare it with the replacement. Make sure the replacement is the same length and will fasten properly.

   Instructor's check _____

6. Install the new link and torque the fasteners to specifications.

   Torque specifications _____

   Instructor's check _____

7. To replaced sway bar bushings, remove the bracket supporting the sway bar and bushing.

   Instructor's check _____

8. Compare the new bushing with the old. Make sure the replacement is the same size and shape and will fit within the bracket.

   Instructor's check _____

9. Install the new bushing and torque the bracket's fasteners to specifications.

   Torque specifications _____

   Instructor's check _____

10. Once finished, put away all tools and equipment and clean up the work area.

   Instructor's check _____

Lab Worksheet 7-10

Name _____ Station _____ Date _____

Inspect and Service Struts

ASE Education Foundation Correlation

This lab worksheet addresses the following **MLR** task:

4.B.16. Inspect, remove, and/or replace strut cartridge or assembly; inspect mounts and bushings. **(P-2)**

4.B.17. Inspect front strut bearing and mount. **(P-1)**

Description of Vehicle

Year _____ Make _____ Model _____

Engine _____ AT MT CVT PSD (circle which applies)

Procedure

1. Refer to the manufacturer's service information for inspecting the struts and strut components on this vehicle. Summarize your findings. _____
_____
_____

2. Perform a visual inspection of the strut assembly. Check for oil leaks around the shock piston. Note your results._____
_____
_____

3. Have an assistant turn the steering wheel while you watch and listen at the strut mount assembly. Watch for looseness or play and listen for any popping, clunking, or grinding noises.

   Summarize your findings. _____
_____
_____

4. If noise or play is found in the strut mount, you will need to remove the strut from the vehicle for further inspection. Refer to the manufacturer's service information about how to remove the strut. Summarize the procedure here. _____
_____
_____

5. Following the service information, remove the strut from the vehicle.

Instructor's check _____

6. Install a spring compressor or mount the strut in a spring compressor and carefully begin to compress the spring. Have your instructor check your work before continuing.

Instructor's check _____

7. Compress the spring until you are able to safely remove the strut piston-to-mount nut. Remove the nut and strut mount. Inspect the mount and bearing. Check for looseness, play, and roughness. Note your findings.

_____

_____

8. Check the shock by pushing the piston down and then extending it back out. There should be resistance to piston movement in both directions. Note your findings.

_____

_____

9. Based on your inspection, what is the recommended action for this strut assembly?

_____

_____

10. Reassemble the strut. Make sure all components are properly aligned. Install the piston-to-mount nut.

Instructor's check _____

11. Install the strut assembly back on the vehicle. Torque all fasteners to specifications.

Torque specifications _____

Instructor's check _____

12. With vehicle back on the ground, jounce the front end and listen for any noise from the strut. Turn the steering wheel from lock to lock and listen for any noise from the bearing and mount. Note your findings.

_____

13. Once finished, put away all tools and equipment and clean up the work area.

Instructor's check _____

Lab Worksheet 7-11

Name _____ Station _____ Date _____

Inspect Rear Suspension

ASE Education Foundation Correlation

This lab worksheet addresses the following **MLR** task:

4.B.18. Inspect rear suspension system lateral links/arms (track bars), control (trailing) arms. **(P-1)**

4.B.19. Inspect rear suspension system leaf spring(s), spring insulators (silencers), shackles, brackets, bushings, center pins/bolts, and mounts. **(P-1)**

Description of Vehicle

Year _____ Make _____ Model _____

Engine _____ AT MT CVT PSD (circle which applies)

Procedure

1. Refer to the manufacturer's service information for inspecting the rear suspension components on this vehicle. Summarize your findings. _____
   _____
   _____

2. Perform a visual inspection of the rear suspension. Check for oil leaks around the shock pistons. Note your results._____
   _____

3. Inspect the bushings for the track bar, control arms, and leaf spring brackets. Note your findings.
   _____
   _____

4. What types of concerns can be caused by worn or damaged suspension bushings?
   _____
   _____
   _____

5. Inspect the leaf springs. Look for broken springs, damage insulators, and hanger. Check the condition of the spring clamps, U-bolts, and leaves. Note your findings. _____

_____

_____

6. Worn or damaged centering pins can cause the rear axle to become misaligned with the leaf springs. Inspect where the springs mount to the axle. Note any signs that the rear axle has shifted in relation to the springs. Note your findings. _____

_____

_____

7. Lower the vehicle and check the rear ride height. Locate the ride height specifications and testing procedures using the service information. Note your findings. _____

_____

_____

8. Based on your inspection, what is the recommended action for this vehicle? _____

_____

_____

9. Once finished, put away all tools and equipment and clean up the work area.

Instructor's check _____

Lab Worksheet 7-12

Name _____ Station _____ Date _____

Inspect Shocks

ASE Education Foundation Correlation

This lab worksheet addresses the following **MLR** task:

4.B.20. Inspect, remove, and/or replace shock absorbers; inspect mounts and bushings. **(P-1)**

Description of Vehicle

Year _____ Make _____ Model _____

Engine _____ AT MT CVT PSD (circle which applies)

Front suspension type _____ Rear suspension type _____

Procedure

Inspect the shocks and struts for oil leaks, noise, mounting hardware, and excessive bounce.

1. Visual inspection

   RF _____
   LF _____
   RR _____
   LR _____

2. Bounce each corner of the vehicle while listening for noises such as creaking and knocking. Note your findings._____

3. If noise is heard, how can the shock and strut be isolated to determine if it is the cause of the noise? ____
   _____

4. Bounce each corner of the vehicle four or five times and count the number of times the vehicle bounces until it stops.

   RF _____
   LF _____
   RR _____
   LR _____

   Based on this test, what condition are the shock and struts? _____
   _____

5. Inspect the shock mountings and bushings for signs of wear. Note your findings. _____
_____

6. Based on your inspection of the shocks and struts, what are your recommendations? _____
_____

7. Remove any tools and equipment and clean up the work area. _____

Instructor's check _____

Lab Worksheet 7-13

Name _____ Station _____ Date _____

Replace Shocks

ASE Education Foundation Correlation

This lab worksheet addresses the following **MLR** task:

4.B.20. Inspect, remove, and/or replace shock absorbers; inspect mounts and bushings. **(P-1)**

Description of Vehicle

Year _____ Make _____ Model _____

Engine _____ AT MT CVT PSD (circle which applies)

Front suspension type _____ Rear suspension type _____

Procedure

Inspect the shocks and struts for oil leaks, noise, mounting hardware, and excessive bounce.

1. Refer to the manufacturer's service information for specific procedures to replace the shock absorbers on this vehicle. Summarize the procedures. _____

   _____

   _____

2. Locate and record and specific safety procedures that apply to replacing the shock absorbers._____

   _____

   _____

3. Safely raise and support the vehicle. Note: Depending on the vehicle, you may need a floor or axle jack to support certain components when replacing the shock.

   Instructor's check _____

4. Remove the shock mounting hardware and remove the shock.

   Instructor's check _____

5. Compare the old and new shock and ensure the replacement shock is correct.

   Instructor's check _____

6. Install the new shock and torque all fasteners to specifications.

   Torque specifications. _____

   Instructor's check _____

7. Remove any supplemental supports and lower the vehicle.

   Instructor's check _____

8. Bounce the vehicle at each corner where a new shock is installed several times. Make sure there are no unusual noises from the shock.

   Instructor's check _____

9. Refer to the manufacturer's service information for proper shock disposal procedures. Note any special procedures here.

   _____

   _____

10. Remove any tools and equipment and clean up the work area.

    Instructor's check _____

Lab Worksheet 7-14

Name _____ Station _____ Date _____

Replace Shocks

ASE Education Foundation Correlation

This lab worksheet addresses the following **MLR** task:

4.C.2. Describe alignment angles (camber, caster, and toe). **(P-1)**

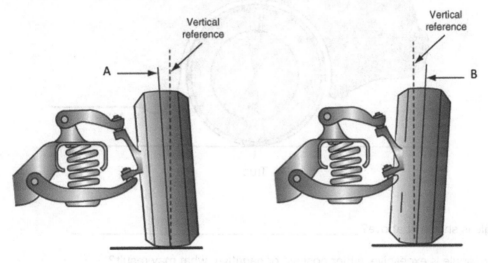

1. Which angle is shown at Point A _____ Point B _____

2. If the above angle is excessive, either positive or negative, what may result?_____
   _____
   _____

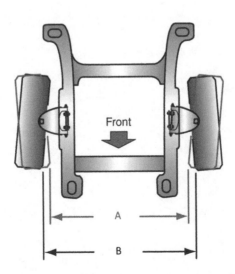

3. Which angle is shown at Point A _____ Point B _____

4. If the above angle is excessive, either positive or negative, what may result? _____

_____

_____

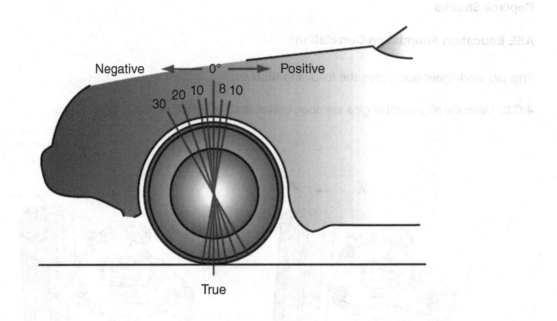

5. Which angle is shown at above? _____

6. If the above angle is excessive, either positive or negative, what may result? _____

_____

_____

Lab Worksheet 7-15

Name _____ Station _____ Date _____

Disable and Enable SRS

ASE Education Foundation Correlation

This lab worksheet addresses the following **MLR** task:

6.E.4. Disable and enable supplemental restraint system (SRS); verify indicator lamp operation. **(P-1)**

Description of Vehicle

Year _____ Make _____ Model _____

Engine _____ AT MT CVT PSD (circle which applies)

Procedure

1. Locate and record the manufacturer's service procedures for disarming the air bag system._____
   _____
   _____

2. If a wait time is specified after disarming the system but before work begins, why is it important for you to follow this step? _____
   _____

3. If the service procedure includes disconnecting the battery, what other systems and components will be affected? _____
   _____

4. What items may need to be reset after the battery is disconnected and reconnected? _____
   _____
   _____

5. Perform the procedure to disarm the air bag.

   Instructor's check _____

6. Re-enable the air bag.

   Instructor's check _____

7. Turn the ignition on and note the SRS warning lamp. Does the lamp illuminate?

Yes _____   No _____

8. If yes, start the engine and note the lamp after several seconds. Does the lamp go out?

Yes _____   No _____

If yes, what does this indicate? _____

9. If the lamp did not illuminate in step 8, recheck your work to ensure the system has been properly re-enabled.

Did you find and correct the problem?   Yes _____   No _____

Instructor's check _____

10. Remove any tools and equipment and clean up the work area.

Instructor's check _____

Lab Worksheet 7-16

Name _____ Station _____ Date _____

Describe Suspension and Steering Components

ASE Education Foundation Correlation

This lab worksheet addresses the following **MLR** task:

4.B.23 Describe the function of suspension and steering control systems and components, (i.e., active suspension and stability control). **(P-3)**

Description of Vehicle

Year _____ Make _____ Model _____

Engine _____ AT MT CVT PSD (circle which applies)

Procedure

1. Using the manufacturer's service information, describe the type of active suspension, traction control, and stability control system used on this vehicle. _____

   _____

   _____

2. List the inputs used by the systems described above. _____

   _____

   _____

3. What concerns may result from a fault in the active suspension or stability control systems? _____

   _____

   _____

4. Describe the purpose of each of these systems:

   Active suspension _____

   _____

   _____

Traction control _____

_____

_____

Stablilty control _____

_____

_____

CHAPTER 8

Steering System Principles

Review Questions

1. The _____ system works with components of the suspension to provide for the turning movement of the wheels.

2. List three functions of the steering system.

    a. _____

    b. _____

    c. _____

3. Leverage, also called _____, is used at the steering wheel and the gearbox to increase force supplied by the driver.

4. The steering column enables the driver to control the direction of the vehicle and provides some _____ to make steering a little easier.

5. Define *steering ratio*. _____
   _____
   _____

6. *Technician A* says all power steering systems use the same power steering fluid. *Technician B* says automatic transmission fluid may be used in place of power steering fluid in modern vehicles. Who is correct?

    a. Technician A

    b. Technician B

    c. Both A and B

    d. Neither A nor B

7. Modern power steering systems use which of the following to supply power assist?

    a. Electricity

    b. Hydraulics

    c. Pneumatics

    d. Both a and b

8. Hydraulic power steering uses a _____ hydraulic power steering pump.

219

9. Explain how tires affect steering and how the vehicle handles. _____

_____

10. List five functions of the steering column in addition to its purpose for steering the vehicle. _____

_____

_____

_____

11. *Technician A* says all collapsible steering columns use the plastic mesh construction that allows the column to collapse. *Technician B* says collapsible steering columns are important safety features in modern cars and trucks. Who is correct?

    a. Technician A

    b. Technician B

    c. Both A and B

    d. Neither A nor B

12. Another method used to prevent injury from the steering assembly during a collision is using a _____ steering shaft.

13. Which of the following components maintains the airbag connection through the steering column?

    a. Impact sensor

    b. Clock spring

    c. Airbag coupler

    d. None of the above

14. The _____ function allows the driver to adjust steering wheel position to increase comfort while driving.

15. A _____ steering wheel can move closer or farther from the driver's seat.

16. The gearbox uses two gears to convert _____ motion of the steering wheel into a _____ motion that moves the wheels.

17. Identify the components of the rack-and-pinion gearbox shown by unlabeled arrows in Figure 8-1.

18. Describe how power assist is obtained through the pinion gear of a power-assisted rack-and-pinion gearbox.

_____

_____

_____

19. Describe the purpose of the ball bearings in the recirculating ball gearbox.

_____

_____

20. Label the power-assist components of the recirculating ball gearbox shown in Figure 8-2.

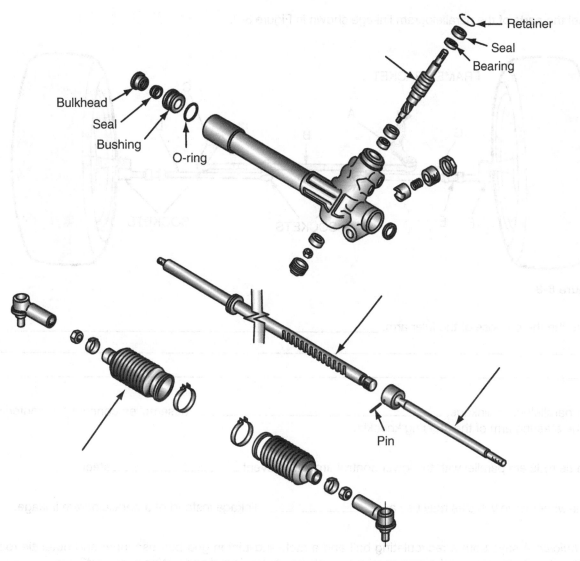

Figure 8-1

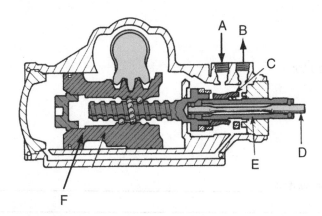

Figure 8-2

21. Label the parts of the parallelogram linkage shown in Figure 8-3.

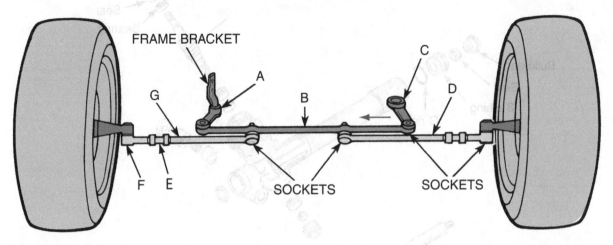

Figure 8-3

22. Describe the purpose of the idler arm. _____
_____
_____

23. In a parallelogram linkage, _____ assemblies connect the centerlink to the steering arm of the steering knuckle.

24. The tie rods are parallel with the lower control arms to prevent _____ steer.

25. Four-wheel-drive vehicles may use a _____ linkage instead of a parallelogram linkage.

26. *Technician A* says both a recirculating ball and a rack-and-pinion gearbox use inner and outer tie rods. *Technician B* says rack-and-pinion gearboxes only use outer tie rod ends. Who is correct?
    a. Technician A
    b. Technician B
    c. Both A and B
    d. Neither A nor B

27. The check valve in a power steering pump is used to:
    a. Limit maximum pressure
    b. Control fluid flow at idle
    c. Boost fluid pressure at high speed
    d. Bypass fluid flow at idle

28. Explain the purpose of a power steering pressure switch. _____
_____
_____

29. A hydraulic power-assist system uses a pump, a high-pressure hose, and a low-pressure _____ hose.

30. A worn and/or loose power steering belt can cause all of the following except:
    a. Erratic power assist
    b. Excessive power assist
    c. Noise
    d. Increased steering effort

31. Electric power assist can be located:
    a. As part of the rack-and-pinion
    b. As part of the ball nut
    c. In the steering column
    d. Both a and c

32. Explain two advantages of electric power steering assist. _____
    _____
    _____

33. Describe how EPS is integrated into other safety and driving systems. _____
    _____
    _____
    _____

Activities

1. Identify the steering components shown in Figure 8-4 (a through d).

a.

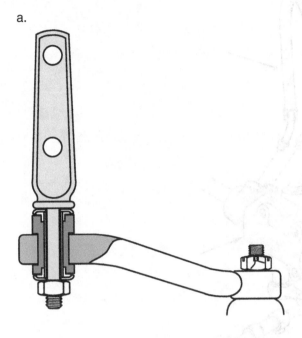

b.

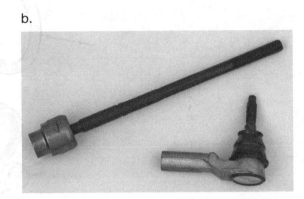

c.

d.

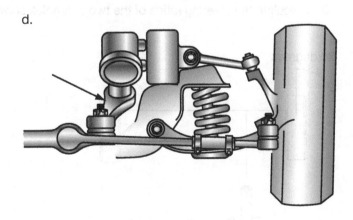

Figure 8-4

2. Label the components of the power steering system shown in Figure 8-5.

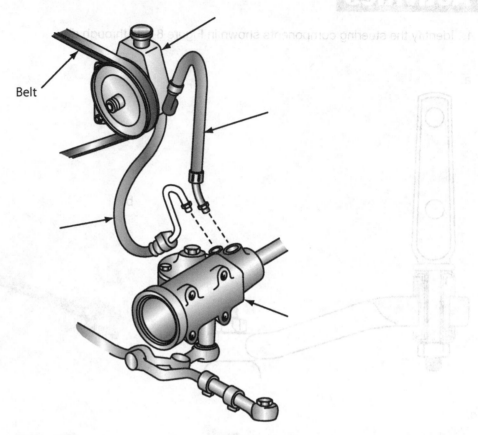

Belt

Figure 8-5

3. Calculate the steering ratios of the two examples shown in Figure 8-6 (a and b).

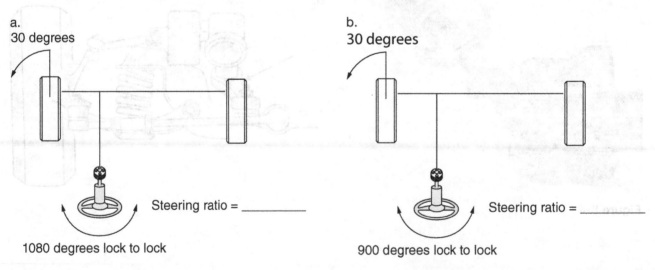

a.
30 degrees

Steering ratio = _____

1080 degrees lock to lock

b.
30 degrees

Steering ratio = _____

900 degrees lock to lock

Figure 8-6

a. Which of the two systems will provide easier steering? _____

_____

b. Which of the two systems will have more road feel? _____

_____

c. Explain why different steering ratios are used for different vehicles. _____

_____

_____

Lab Activity 8-1

Name _____ Date _____ Instructor _____

Year _____ Make _____ Model _____

Identify steering linkage types and components

1. Steering gearbox type _____

2. Type of steering linkage (circle one)   Parallelogram   Rack-and-pinion

   Haltenburger   Crosslink   Other

3. Components used (circle all that apply)   Inner tie rods   Outer tie rods

   Tie rod sleeves   Centerlink   Pitman arm   Idler arm   Drag link

   Year _____ Make _____ Model _____

1. Steering gearbox type _____

2. Type of steering linkage (circle one)   Parallelogram   Rack-and-pinion

   Haltenburger   Crosslink   Other

3. Components used (circle all that apply)   Inner tie rods   Outer tie rods

   Tie rod sleeves   Centerlink   Pitman arm   Idler arm   Drag link

   Year _____ Make _____ Model _____

1. Steering gearbox type

2. Type of steering linkage (circle one) Parallelogram   Rack-and-pinion

   Haltenburger   Crosslink   Other

3. Components used (circle all that apply)   Inner tie rods   Outer tie rods

   Tie rod sleeves   Centerlink   Pitman arm   Idler arm   Drag link

Lab Activity 8-2

Name _____ Date _____ Instructor _____

Year _____ Make _____ Model _____

Steering column inspection

1. Steering column components

    Tilt _____ Telescoping _____

    Memory _____ Electric assist _____

2. Operate the tilt and telescoping functions (if applicable). Note any of the following.

    a. Binding, tightness, looseness _____ Yes _____ No _____

    b. Noises _____ Yes _____ No _____

    If you checked Yes, describe the problem and when it occurs. _____

    _____

    _____

3. With the engine off and the steering unlocked, turn the steering wheel from lock to lock. Note any noises from the column. _____

4. Start the engine and turn the wheel from lock to lock. Note any indication of binding or looseness in the column. _____

5. Based on your inspection, describe the condition of the steering column. _____

    _____

CHAPTER 9

Steering Service

Review Questions

1. To remove a _____ arm, a puller is required because of the tapered fit between the arm and the gearbox.

2. Follow the manufacturer's service procedures exactly when working on the _____ system.

3. Removing the airbag or SRS _____ is a common step when disarming the airbag system.

4. Explain why removing the battery cable is not always recommended as part of the airbag disarming process.
   _____
   _____
   _____

5. Spending time gathering _____ from the customer can often save time in the actual diagnosis.

6. *Technician A* says problems with the suspension system components cannot affect the steering system. *Technician B* says the two systems share some components and both systems should be inspected. Who is correct?
   a. Technician A
   b. Technician B
   c. Both A and B
   d. Neither A nor B

7. Describe how to check power steering fluid level and condition. _____
   _____
   _____
   _____

8. Describe how to remove an inner tie rod from an end-takeoff rack and pinion._____

_____

_____

_____

9. The tool shown in Figure 9-1 is used to:
   a. Remove the outer tie rod end
   b. Remove the rack-and-pinion boot
   c. Torque the castle nut
   d. Remove and install the inner rack tie rod

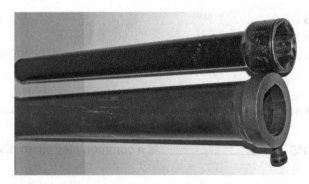

Figure 9-1

10. When a tie rod has been replaced, the _____ angle setting should be checked and if necessary, adjusted to specs.

11. When installing a new component that uses a castle nut, *Technician A* tightens the nut to specs and then loosens it until the cotter pin hole aligns with the nut. *Technician B* tightens the nut to specs and then continues to tighten the nut until the cotter pin hole aligns. Who is correct?
    a. Technician A
    b. Technician B
    c. Both A and B
    d. Neither A nor B

12. Vehicles with recirculating ball gearboxes, such as those with SLA suspensions, usually use the _____ steering linkage.

13. Centerlinks and Pitman arms can be either wear or _____ types.

14. If nylon friction nuts are used instead of castle nuts, they should be _____ once removed.

15. *Technician A* says it is OK to reuse an old cotter pin if it is still in good condition. *Technician B* always replaces old cotter pins with new. Who is correct?

    a. Technician A

    b. Technician B

    c. Both A and B

    d. Neither A nor B

16. The tool shown in Figure 9-2 is used to:

    a. install tie rod ends.

    b. remove Pitman arms.

    c. install idler arms.

    d. none of the above.

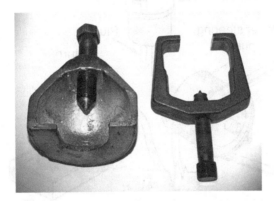

Figure 9-2

17. *Technician A* says idler arms should have zero play in the sockets. *Technician B* says some idler arms are allowed a slight amount of play. Who is correct?

    a. Technician A

    b. Technician B

    c. Both A and B

    d. Neither A nor B

18. A steering damper is similar to a _____ .

19. Which of the following is least likely to be the cause of a hard steering complaint?

    a. Defective power steering pump

    b. Worn power steering gearbox

    c. Worn bearings in the steering column

    d. Low power steering fluid level

20. Electric power assist can be applied either at the _____ or to the rack and pinion.

21. Technician A says all modern vehicles can use either ATF or power steering fluid in the power steering system. Technician B says some vehicle manufacturers require special power steering fluid in their systems. Who is correct?

    a. Technician A

    b. Technician B

    c. Both A and B

    d. Neither A nor B

22. Explain what service is being performed in Figure 9-3. _____

    _____

    _____

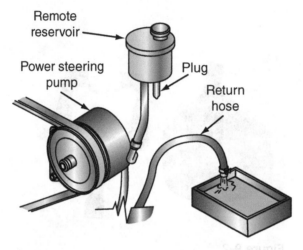

Figure 9-3

23. When flushing and bleeding a power steering system, why should the steering be held in the full-turn lock position for more than a few seconds? _____

    _____

    _____

24. Some power steering systems use a _____ inline with a power steering hose or in the reservoir.

25. Power steering hoses wear from the _____ out.

26. When a leaking hose is being replaced, it is a good idea to replace _____ hoses.

27. Describe how to replace a power steering hose. _____

    _____

    _____

    _____

28. Power steering belts should be _____ during routine service and often require _____ every 60,000 to 90,000 miles.

29. Describe three ways in which tension is applied to a power steering belt._____
_____
_____

30. Poor power steering performance may be caused by all of the following except:
    a. Loose power steering belt
    b. Worn belt tensioner
    c. Binding steering coupler
    d. Low power steering fluid level

31. Some vehicles use a combination of _____ and _____ to provide variable power assist.

32. Some vehicles use electric _____ to provide all of the power steering assist.

33. Begin inspecting the electric power-assist system by noting if any _____ lights are illuminated on the dash.

34. A fault in the electric power assist may result in the system setting a _____ _____ code.

35. Components of the _____-_____ system on a hybrid vehicle are bright orange.

Activities

1. Make a list of steering and suspension components that can cause hard steering.

_____

_____

_____

_____

_____

Lab Worksheet 9-1

Name _____ Station _____ Date _____

Disable and Enable SRS

ASE Education Foundation Correlation

This lab worksheet addresses the following **MLR** task:

4.A.2. Disable and enable supplemental restraint system (SRS); verify indicator lamp operation. **(P-1)**

Description of Vehicle

Year _____ Make _____ Model _____

Engine _____ AT MT CVT PSD (circle which applies)

Procedure

1. Locate and record the manufacturer's service procedures for disarming the air bag system. _____

   _____

   _____

2. If a wait time is specified after disarming the system but before work begins, why is it important for you to follow this step? _____

   _____

   _____

3. If the service procedure includes disconnecting the battery, what other systems and components will be affected? _____

   _____

   _____

4. What items may need to be reset after the battery is disconnected and reconnected? _____

   _____

   _____

5. Perform the procedure to disarm the air bag.

   Instructor's check _____

6. Re-enable the air bag.

   Instructor's check _____

7. Turn the ignition on and note the SRS warning lamp. Does the lamp illuminate?

   Yes _____No _____

8. If yes, start the engine and note the lamp after several seconds. Does the lamp go out? _____

   Yes _____No _____

   If yes, what does this indicate? _____

9. If the lamp did not illuminate in step 8, recheck your work to ensure the system has been properly re-enabled.

   Did you find and correct the problem?   Yes _____   No _____

                                                        Instructor's check _____

10. Remove any tools and equipment and clean up the work area.

                                                        Instructor's check _____

Lab Worksheet 9-2

Name _____ Date _____ Instructor _____

Inspect Inner Tie Rods

ASE Education Foundation Correlation

This lab worksheet addresses the following **MLR** task:

4.B.1. Inspect rack and pinion steering gear inner tie rod ends (sockets) and bellows boots. **(P-1)**

Description of Vehicle

Year _____ Make _____ Model _____

Engine _____ AT MT CVT PSD (circle which applies)

Procedure

1. Refer to the manufacturer's service information for inspecting steering system components. Summarize your findings. _____

   _____

   _____

2. To perform a dry-park check of the steering linkage: place the vehicle on a drive-on lift. Have an assistant shake the steering wheel back and forth rapidly from center with the engine off. While your assistant is shaking the steering wheel, inspect for noise and looseness at the inner tie rod connections and bellows. A loose socket will often move in and out and often cause a knocking sound. Place your hand on the bellows to feel any looseness in the socket. Note your findings. _____

   _____

   _____

3. To manually check the steering linkage, grasp each component near its socket and attempt to move the component up and down and side to side. A worn and loose socket will usually have noticeable play and movement. Note your findings. _____

   _____

   _____

4. Inspect the bellows for signs of damage and power steering fluid loss. Note your findings. _____

   _____

   _____

5. If fluid is leaking from the bellows, what does this indicate? _____

_____

6. Describe what types of concerns can be caused by loose or damaged steering linkage components. _____

_____

_____

7. Based on your inspection, what are the necessary actions for this vehicle? _____

_____

_____

8. Once finished, put away all tools and equipment and clean up the work area.

Instructor's check _____

Lab Worksheet 9-3

Name _____ Date _____ Instructor _____

Inspect Power Steering Fluid.

ASE Education Foundation Correlation

This lab worksheet addresses the following **MLR** task:

4.B.2. Inspect power steering fluid level and condition. **(P-1)**

Description of Vehicle

Year _____ Make _____ Model _____

Engine _____ AT MT CVT PSD (circle which applies)

Procedure

1. Locate and record the manufacturer's recommended power steering fluid. _____

2. Determine the location of the power steering fluid reservoir.

   Remote reservoir _____   Pump and reservoir _____

3. If the fluid level can be seen through the reservoir. Note the level. _____

4. If the cap must be removed to check the level, clean the cap and reservoir around the cap before removing. Remove the cap and note the level on the dipstick. _____

5. Note the color of the fluid. _____

6. Remove a few drops of fluid and wipe on a clean, light-colored shop towel. Inspect the fluid for metal shavings, color, and smell. Note your findings. _____

7. Based on your inspection, what is the condition of the fluid? _____

8. Remove any tools and equipment and clean up the work area.

   Instructor's check _____

Lab Worksheet 9-4

Name _____ Date _____ Instructor _____

P/S Fluid Service

ASE Education Foundation Correlation

This lab worksheet addresses the following **MLR** tasks:

4.B.3. Flush, fill, and bleed power steering system; use proper fluid type per manufacturer specification. **(P-2)**

Description of Vehicle

Year _____ Make _____ Model _____

Engine _____ AT MT CVT PSD (circle which applies)

Procedure

Refer to the manufacturer's service information for flushing the power steering system. Summarize the procedure. _____

_____

_____

If there is not a specific procedure in the service information, follow these steps.

1. Does this power steering system use a remotely mounted reservoir? Yes_____No _____

2. If yes, place a shop pan under the reservoir and remove the power steering return hose from the power steering fluid reservoir. If the reservoir is integral to the pump, remove the return line at the pump.

3. Cap the opening of the pump or reservoir from which the return line was attached. Method of capping the opening _____

4. Determine and note the correct power steering fluid type. _____

Note: Using the wrong fluid can cause damage to the power steering system.

5. Fill the reservoir with new power steering fluid. Have additional fluid ready. _____

Instructor's check _____

6. Have an assistant start the engine and turn the steering wheel from lock to lock as you add fluid to the reservoir. Continue to flush the system until the old fluid is gone and new fluid is coming out of the return line.

Instructor's check _____

7. Uncap the pump or reservoir and reattach the return line.

Instructor's check _____

8. Top off the power steering reservoir with new fluid. Start the engine and turn the steering wheel from lock to lock several times to bleed any trapped air from the system.

Instructor's check _____

9. Air trapped in the system can cause what problems? _____

_____

_____

10. With the engine off and the wheels straight ahead, check the power steering fluid level. Top off as needed.

Instructor's check _____

11. Why is it important to not overfill the power steering system? _____

_____

_____

12. Once complete, clean up and spilled fluid. Clean the vehicle and make sure the power steering fluid cap is on securely.

Instructor's check _____

13. Once finished, put away all tools and equipment and clean up the work area.

Instructor's check _____

Lab Worksheet 9-5

Name _____ Date _____ Instructor _____

P/S Leaks

ASE Education Foundation Correlation

This lab worksheet addresses the following **MLR** task:

4.B.4. Inspect for power steering fluid leakage. **(P-1)**

Description of Vehicle

Year _____ Make _____ Model _____

Engine _____ AT MT CVT PSD (circle which applies)

Procedure

1. Check and record power steering fluid level. _____

2. Inspect the following locations and note any signs of fluid leakage.

Power steering pump	OK _____	Not OK _____
Power steering pressure hose	OK _____	Not OK _____
Power steering return hose	OK _____	Not OK _____
Power steering pressure switch	OK _____	Not OK _____
Pinion seals and worm shaft seals	OK _____	Not OK _____
Rack fluid transfer lines	OK _____	Not OK _____
Rack lower pinion seals	OK _____	Not OK _____
Pitman shaft seals	OK _____	Not OK _____
Rack end seals (bellows)	OK _____	Not OK _____
Power steering cooler lines	OK _____	Not OK _____
Power steering cooler	OK _____	Not OK _____

3. Based on your inspection, what is the necessary action?

_____

_____

_____

4. Remove any tools and equipment and clean up the work area.

Instructor's check _____

Lab Worksheet 9-6

Name _____ Date _____ Instructor _____

Replace a Power Steering Belt

ASE Education Foundation Correlation

This lab worksheet addresses the following **MLR** task:

4.B.5. Remove, inspect, replace, and/or adjust power steering pump drive belt. **(P-1)**

Description of Vehicle

Year _____ Make _____ Model _____

Engine _____ AT MT CVT PSD (circle which applies)

Procedure

1. Locate the power steering pump belt. Belt type _____
   _____

2. Describe how the belt is loosened or tension is removed. _____
   _____

3. What tools are required to remove the belt? _____
   _____

4. Do any other belts need to be removed to remove the power steering belt? _____

   Yes _____   No _____

   If yes, which belt(s)? _____

5. Remove the power steering belt and note its condition. _____
   _____

   Instructor's check _____

6. Locate the manufacturer's specifications for proper belt tension and record. _____

7. Install the belt and apply tension. Check for proper tension with a gauge and record your findings.

_____

_____

Instructor's check _____

8. Make sure the belt is properly aligned and seated in all pulleys. Start the engine and confirm proper operation.

Instructor's check _____

9. Remove any tools and equipment and clean up the work area.

Instructor's check _____

Lab Worksheet 9-7

Name _____ Date _____ Instructor _____

Replace P/S Hoses

ASE Education Foundation Correlation

This lab worksheet addresses the following **MLR** task:

4.B.6. Inspect and replace power steering hoses and fittings. **(P-2)**

Description of Vehicle

Year _____ Make _____ Model _____

Engine _____ AT MT CVT PSD (circle which applies)

Procedure

1. Inspect the vehicle for a power steering fluid leak. Note your findings. _____
   _____
   _____

2. If the leak is from a power steering hose, what is the procedure to replace the hose?
   _____
   _____
   _____

   Instructor's check _____

3. Place a shop pan under the area where you will be working. Loosen the hose fittings and remove any clamps or brackets holding the hose in place. Remove the hose.

   Instructor's check _____

4. Compare the new and old hose. Make sure the fittings are the same size and thread. Check the length of the hose and transfer any brackets or other hardware as necessary.

   Instructor's check _____

5. Place the new hose in position and carefully begin to thread the fittings in place. Be careful not to cross-thread the fittings. Once you have the fittings started, seat the fittings into place.

   Instructor's check _____

6. Locate and record the fitting torque specifications.

Torque specifications _____

7. Tighten the fittings to specifications.

Instructor's check _____

8. Why is it important to tighten the fittings to specifications? _____
_____
_____

9. Once the line is installed, refill the system with the correct power steering fluid.

Fluid used _____

10. Start the engine and bleed the air from the system. How is air bled from the system? _____
_____
_____

11. Shut the engine off and perform a final check for leaks from the hose. Top off the fluid as necessary.

Instructor's check _____

12. Once finished, put away all tools and equipment and clean up the work area.

Instructor's check _____

Lab Worksheet 9-8

Name _____ Date _____ Instructor _____

Inspect Linkage Components

ASE Education Foundation Correlation

This lab worksheet addresses the following **MLR** tasks:

4.B.7. Inspect pitman arm, relay (centerlink/intermediate) rod, idler arm, mountings, and steering linkage damper. **(P-1)**

4.B.8. Inspect tie rod ends (sockets), tie rod sleeves, and clamps. **(P-1)**

Description of Vehicle

Year _____ Make _____ Model _____

Engine _____ AT MT CVT PSD (circle which applies)

Procedure

1. Refer to the manufacturer's service information for inspecting steering system components. Summarize your findings. _____
_____
_____

2. To perform a dry-park check of the steering linkage: place the vehicle on a drive-on lift. Have an assistant shake the steering wheel back and forth rapidly from center with the engine off. While your assistant is shaking the steering wheel, look at each steering linkage component and connection. A loose socket will often move up and down or move further than the component it is connect to. Place your hand on each socket to feel any looseness in the socket. Note your findings. _____
_____
_____

3. To manually check the steering linkage, grasp each component near its socket and attempt to move the component up and down and side to side. A worn and loose socket will usually have noticeable play and movement. Note your findings. _____
_____
_____

4. Inspect the steering damper for signs of oil loss. Check the damper's mounting bushings for deterioration and damage. Note your findings. _____
_____

5. Describe what types of concerns can be caused by loose or damaged steering linkage components.

_____

_____

_____

6. Based on your inspection, what are the necessary actions for this vehicle?

_____

_____

_____

7. Once finished, put away all tools and equipment and clean up the work area.

Instructor's check _____

Lab Worksheet 9-9

Name _____ Date _____ Instructor _____

Inspect Tie Rods

ASE Education Foundation Correlation

This lab worksheet addresses the following **MLR** task:

4.B.8. Inspect tie rod ends (sockets), tie rod sleeves, and clamps. **(P-1))**

Description of Vehicle

Year _____ Make _____ Model _____

Engine _____ AT MT CVT PSD (circle which applies)

Procedure

1. Refer to the manufacturer's service information for inspecting steering system components. Summarize your findings. _____
   _____
   _____

2. To perform a dry-park check of the steering linkage: place the vehicle on a drive-on lift. Have an assistant shake the steering wheel back and forth rapidly from center with the engine off. While your assistant is shaking the steering wheel, inspect for noise and looseness at the inner tie rod connections and bellows. A loose socket will often move in and out and often cause a knocking sound. Place your hand on the bellows to feel any looseness in the socket. Note your findings. _____
   _____
   _____

3. To manually check the steering linkage, grasp each component near its socket and attempt to move the component up and down and side-to-side. A worn and loose socket will usually have noticeable play and movement. Note your findings. _____
   _____
   _____

4. Inspect the tie rod sleeves and clamps for damage and rust-through. Describe your findings. _____
   _____
   _____

5. Based on your inspection, what are the necessary actions for this vehicle? _____

_____

_____

6. Once finished, put away all tools and equipment and clean up the work area.

Instructor's check _____

Lab Worksheet 9-10

Name _____ Date _____ Instructor _____

Inspect EPS

ASE Education Foundation Correlation

This lab worksheet addresses the following **MLR** task:

4.B.21. Inspect electric power steering assist system. **(P-2)**

Description of Vehicle

Year _____ Make _____ Model _____

Engine _____ AT MT CVT PSD (circle which applies)

Procedure

1. Turn the ignition on and note the EPS warning light on the dash. Does the light illuminate at key-on bulb check? Yes _____ No _____

   If not, what might this indicate? _____

2. Start the engine and note the EPS light.

   Light goes off _____

   Light stays on _____

   Light blinks _____

   What does this indicate? _____

3. Connect a scan tool to the diagnostic link connector (DLC) and navigate to the EPS menu. Check for stored diagnostic trouble codes (DTCs). Note any stored or history DTCs. _____

   _____

4. Navigate to the EPS data menu. With the engine running and the steering wheel straight ahead, note the following data.

Steering shaft torque _____

Steering motor temperature _____

Torque sensor V/N m _____

Steering angle sensor degrees _____

Motor amperage _____

5. With the engine running and the steering wheel turned 90 degrees and held, note the following data.

Steering shaft torque _____

Steering motor temperature _____

Torque sensor V/N m _____

Steering angle sensor degrees _____

Motor amperage _____

6. Based on your inspection, note any problems or concerns with the EPS. _____

_____

_____

7. Remove any tools and equipment and clean up the work area.

Instructor's check _____

Lab Worksheet 9-11

Name _____ Date _____ Instructor _____

Identify Hybrid P/S Safety Precautions

ASE Education Foundation Correlation

This lab worksheet addresses the following **MLR** task:

4.B.22. Identify hybrid vehicle power steering system electrical circuits and safety precautions. **(P-2)**

Description of Vehicle

Year _____ Make _____ Model _____

Engine _____ AT MT CVT PSD (circle which applies)

Procedure

1. Refer to the manufacturer's service information for inspecting the electric power steering system on this vehicle. Summarize your findings. _____

   _____

2. Does the vehicle use high-voltage for the power steering system?   Yes _____   No _____

3. If yes, what color are the high-voltage wires and connections? _____

4. If yes, locate any service precautions and warnings in the service information regarding the power steering system. Note your findings. _____

   _____

5. If the power steering system is not operated by the high-voltage system, describe how the system operates. _____

   _____

   _____

6. List any service precautions for inspecting and servicing the electric power steering system.

   _____

   _____

   _____

Lab Worksheet 9-12

Name _____ Date _____ Instructor _____

Prealignment Inspection

ASE Education Foundation Correlation

This lab worksheet addresses the following **MLR** task:

4.C.1. Perform a prealignment inspection; measure vehicle ride height. **(P-1)**

Description of Vehicle

Year _____ Make _____ Model _____

Engine _____ AT MT CVT PSD (circle which applies)

Procedure

1. Tire size _____

   Locate the tire decal on this vehicle and determine if the installed tire is correct according to the decal.

   Correct   Yes _____   No _____

   If not, what is incorrect? _____

2. Inspect the tires. Note your findings. _____

3. Check and set tire pressure to specifications.   Pressure specification _____

   LF pressure _____   RF pressure _____

   RR pressure _____   LR pressure _____

4. Locate and record the ride height specification.   Specification _____

   Measure and record the ride height.   LF _____   RF _____

   RR _____   LR _____

5. Raise and support the vehicle to inspect the suspension and steering components.

Ball joints   OK _____   Not OK _____

Control arms and bushings   OK _____   Not OK _____

Tie rods and linkage   OK _____   Not OK _____

Wheel bearings   OK _____   Not OK _____

6. Based on your inspection, what is the condition of the vehicle? _____

_____

Lab Worksheet 9-13

Name _____ Station _____ Date _____

Service Belts

ASE Education Foundation Correlation

This lab worksheet addresses the following **MLR** task:

1.C.2. Inspect, replace, and adjust drive belts, tensioners, and pulleys; check pulley and belt alignment. **(P-1)**

Description of Vehicle

Year _____ Make _____ Model _____

Engine _____ AT MT CVT PSD (circle which applies)

Procedure

1. Locate and note the type of drive belt(s) used.

    V-belt _____   Serpentine (multirib) belt _____

2. Describe how tension is applied to the drive belt(s). _____
   _____

3. Inspect the drive belt for wear and damage. Note your findings. _____
   _____
   _____

4. Locate the belt tension specifications and record them. Specifications _____

5. Using a belt tension gauge, measure and record drive belt tension. _____
   Measured tension _____

6. How does the measured tension compare to the specification? _____
   _____

7. How can incorrect belt tension affect the belt and the belt-driven accessories? _____

_____

_____

8. Remove the drive belt and inspect the belt and all accessory drive pulleys. Check pulleys for wear, damage, and faulty bearings. Note your findings. _____

_____

_____

9. Based on your inspection, what do you recommend and why? _____

10. Using a belt alignment tool or straightedge, check for proper alignment of all drive pulleys. Note your findings. _____

_____

11. What concerns can result from improper drive pulley alignment? _____

_____

_____

12. Reinstall the old belt or a new belt as necessary. Ensure the belt is installed correctly onto each pulley before starting the engine.

Instructor's check _____

13. With your instructor's permission, start the engine and ensure proper belt operation.

Instructor's check _____

14. Shut the engine off and recheck the belt.

Instructor's check _____

15. Remove any tools and equipment and clean up the work area.

Instructor's check _____

CHAPTER 10

Brake System Principles

Review Questions

1. Modern brake systems combine the principles of _____, _____, and electronics.

2. The _____ of _____ is a number that expresses the ratio of force required to move an object divided by the mass of the object.

3. Define *brake fade* and describe how it can occur. _____
   _____
   _____
   _____

4. Explain why heat dissipation is important for the brake system. _____
   _____
   _____

5. *Technician A* says solid brake rotors can dissipate heat more effectively than vented rotors. *Technician B* says vented brake rotors are standard equipment on the front of all modern cars and trucks. Who is correct?
   a. Technician A
   b. Technician B
   c. Both A and B
   d. Neither A nor B

6. Leverage is also called mechanical _____

7. The brake pedal acts as a _____, increasing the amount of force applied to the master cylinder pushrod.

8. A typical brake pedal may have a leverage ratio of:
   a. 1:1
   b. 3:1
   c. 20:1
   d. 100:1

9. Leverage involves a trade-off: increasing _____ requires moving a greater distance.

10. The slight amount of pedal movement at the released position before the pushrod begins to move into the booster and master cylinder is called:

    a. Pedal travel

    b. Pedal height

    c. Free play

    d. Compliance

11. *Technician A* says the brake light switch may be used as an input for the on-board computer system. *Technician B* says some vehicles use vacuum-operated brake light switches. Who is correct?

    a. Technician A

    b. Technician B

    c. Both A and B

    d. Neither A nor B

12. Some vehicles have _____ pedals that can move based on driver preference.

13. _____ is the science of using liquids to perform work.

14. By pressurizing a fluid in a closed system, the fluid can transmit both _____ and _____.

15. Pascal determined that _____ exerted on a confined liquid caused an increase in pressure at all points.

16. Which of the following statements about hydraulics is incorrect?

    a. Hydraulic pressure in a closed system is constant.

    b. A smaller input piston generates less hydraulic pressure than a larger piston.

    c. If output pressure increases, output piston travel decreases.

    d. If output pressure decreases, output piston travel increases.

17. Describe in your own words how hydraulic principles are used in the hydraulic brake system. _____

_____

_____

_____

_____

18. The hydraulic brake system input cylinder is called the _____.

19. If 500 pounds is applied to a master cylinder piston with two square inches of surface area, the pressure in the cylinder will be:

    a. 250 psi

    b. 500 psi

    c. 1,000 psi

    d. 2,000 psi

20. The front brake hydraulic output is called a disc brake _____

21. Explain why the pistons used in the disc brake caliper are much larger than the pistons in the master cylinder.
_____
_____

22. Explain why the pistons used in rear drum brakes are much smaller than the pistons used in the front disc brakes. _____
_____
_____

23. To operate properly, the hydraulic brake system must be a _____ system, meaning there is no opening for fluid to leak or vent.

24. Why do modern master cylinders have two chambers and pistons? _____
_____
_____

25. Explain why some master cylinders use two different-sized pistons. _____
_____
_____

26. The reservoir cap seals are often _____ seals, meaning that they _____ in size as the fluid level in the reservoir drops.

27. *Technician A* says the master cylinder primary piston is located in the front of the master cylinder. *Technician B* says the master cylinder primary piston is located in the rear of the cylinder. Who is correct?
    a. Technician A
    b. Technician B
    c. Both A and B
    d. Neither A nor B

28. All of the following describe brake line material except:
    a. Flared ends to prevent leaks
    b. Made from corrosion-resistant aluminum
    c. has double wall thickness for strength
    d. metric and standard flares not interchangeable

29. A _____ splits a single brake line so that each rear wheel brake receives brake fluid.

30. Which is the function of the proportioning valve?
    a. Limit pressure to the front disc brakes
    b. Limit pressure to the rear disc brakes
    c. Limit pressure to the rear drum brakes
    d. Illuminate the brake warning light on the dash

31. What is the function of the load-sensing proportioning valve?

    a. Limit pressure to the front disc brakes

    b. Control rear drum brake pressure based on vehicle load

    c. Shut off a brake circuit in the event of a leak

    d. Illuminate the brake warning light on the dash

32. The _____ valve is used to delay slightly the application of the front disc brakes.

33. Which valve is responsible for closing off one of the hydraulic circuits if a leak occurs?

    a. Metering valve

    b. Proportioning valve

    c. Pressure differential valve

    d. None of the above

34. The pressure differential valve usually contains an _____ contact that completes the brake warning lamp circuit.

35. Instead of a vehicle having three separate brake valves, they are commonly all together in a _____ valve.

36. Calipers and wheel cylinders are the _____ of the hydraulic system.

37. A fixed caliper has at least how many pistons?

    a. Two

    b. Three

    c. Four

    d. Six

38. Which caliper component is responsible for retracting the piston when the brakes are released?

    a. Caliper return spring

    b. Caliper dust boot

    c. Square seal

    d. Caliper bracket guide

39. _____ calipers are the most common type of brake caliper used in modern cars and light trucks.

40. Newton's Third Law of Motion states that for every _____ there is an _____ and _____ reaction.

41. Explain how Newton's Third Law of Motion applies to the disc brake system. _____

    _____

    _____

42. List the components of a typical front brake caliper. _____

    _____

    _____

43. Explain the operation of the integral parking brake rear disc brake caliper. _____

_____

_____

44. Some newer vehicles use _____ operated parking brake calipers.

45. Wheel cylinders are used with the _____ brake arrangement.

46. Explain what DOT 3, DOT 4, and DOT 5 brake fluid ratings mean. _____

_____

_____

47. List five qualities important for brake fluid. _____

_____

_____

48. *Technician A* says DOT 3 and DOT 4 are compatible fluids. *Technician B* says DOT 4 should not be used with vehicles equipped with antilock brakes (ABS). Who is correct?

    a. Technician A

    b. Technician B

    c. Both A and B

    d. Neither A nor B

49. If a fluid is _____ that means that it easily absorbs moisture.

50. Why must a technician be careful when working with brake fluid? _____

_____

_____

51. In conventional vehicles when the brakes are applied, the _____ energy of the vehicle is converted into _____ energy at the brakes.

52. Hybrid electric vehicles use the _____ system to recover brake energy to _____ the high-voltage batteries.

Activities

I. Leverage

The brake pedal in a car operates as a second-class lever. With a second-class lever, the fulcrum is at one end of the lever instead of the middle. The effort is applied to the other end of lever, and the force is applied somewhere in between the effort and the fulcrum, like that shown in Figure 10-1. Look under the dashboard of several vehicles to see how the brake pedal is mounted. Notice that the top of the pedal is the mounting point, which acts as the fulcrum. Below the fulcrum, the pedal pushrod is mounted and extends forward through the firewall, to the brake booster, and master cylinder. The footpad at the bottom of the pedal is where the effort is applied.

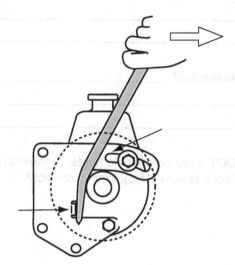

Figure 10-1

Measure the length of the pedal between the pushrod and the fulcrum and the pushrod and the lowest point of the pedal for several vehicles.

1. Pushrod-to-fulcrum length _____ Pushrod-to-pedal length _____

2. Pushrod-to-fulcrum length _____ Pushrod-to-pedal length _____

Because the force is between the fulcrum and the effort, the resulting force is going to be determined by the distance below the force divided by the distance above the force. Use the following formula: $R = d_a/d_b$. R is the ratio, d_a is the distance from the pushrod to the end of pedal, and d_b is the distance from the pushrod to the fulcrum. Using the measurements of the brake pedals from questions 1 and 2, determine the ratio of force for each vehicle.

3. d_a/d_b = _____   Ratio _____

4. d_a/d_b = _____   Ratio _____

As you can see, the brake pedal and its mounting design reduce the amount of effort needed from the driver. This reduction of effort is beneficial for the driver, as it will reduce fatigue over the time of vehicle operation.

II. Hydraulics

Even though the force applied to the brake pedal is increased, it is necessary for the hydraulic system to further increase the force and apply it in a manner that will safely slow and stop the vehicle.

A simple hydraulic system, like that in Figure 10-2, contains two equal-sized containers of a liquid with two equal-sized pistons. The two containers are connected with a hose or tubing. If one piston is pushed downward with 100 lb (45 kg) of force and moves down 1 inch, the piston in the second container will move up 1 inch with the same 100 lb (45 kg) of force. Since the pistons are the same size, any force and movement imparted on one piston will cause the same reaction to the second piston.

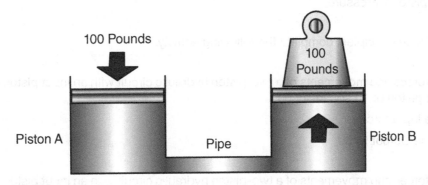

100 Pounds

100 Pounds

Piston A   Pipe   Piston B

Figure 10-2

The use of hydraulics really provides advantage when the sizes of the pistons are different: the resulting force and movement can be increased or decreased as needed. The pressure generated by the piston is a factor of the piston size. Input pressure is found by dividing force by piston size, or $P = F/A$. The smaller the input piston surface area, the larger the force will be from that piston. The larger the input piston surface area, the less the force will be from that piston. Conversely, the output piston force is proportional to the pressure against the surface area of the piston. An output piston that is larger than the input piston will move with greater force than the input piston but over a shorter distance. An example is shown in Figure 10-3. To examine this principle we will look at what is known as Pascal's principle or law.

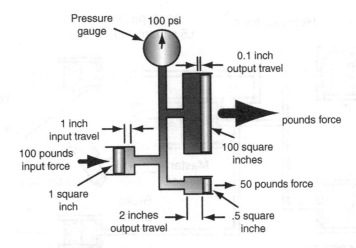

Pressure gauge

100 psi

0.1 inch output travel

pounds force

1 inch input travel

100 pounds input force

100 square inches

1 square inch

50 pounds force

2 inches output travel

.5 square inche

Figure 10-3

If the smaller piston on the left is our input piston and has a surface area of 1 square inch (6.45 cm²) and the larger piston is our output piston and has an area of 10 square inches (65.5 cm²), we can calculate the forces and movements quite easily. Suppose a force of 100 lb (45 kg) is exerted on our input piston, moving it 1 inch (2.54 cm).

Using $P = F/A$, we can calculate $P = 100/1$, or 100 psi. Our output piston, which has a surface area of 10 square inches, will receive 100 pounds of pressure per square inch. This will result in an output force by piston 2 of 1,000 lbs. However, since our output force has increased, our output movement will decrease. The larger piston will move 1/10 of the distance of the input piston; since the force was multiplied by 10, the distance will be divided by 10 as well. The 1 inch of movement of piston 1 turns into 1/10 of an inch of movement at piston 2. To calculate the forces and movement in a hydraulic circuit, use the following formulas:

$P_1 = F_1/A_1$ for piston 1 pressure

$F_2 = A_2/A_1 \cdot F_1$ for piston 2 force

$P_2 = F_2/A_2$ for piston 2 pressure

To practice using these principles, complete the following activity.

1. Calculate the forces and movements of a two-piston hydraulic circuit with an input piston of 2 square inches and an output piston of 10 square inches.

   Input distance Input force

   Output distance Output force

2. Calculate the forces and movements of a two-piston hydraulic circuit with an input piston of 2 square inches and an output piston of 1 square inch.

   Input distance _____ Input force _____

   Output distance _____ Output force _____

   The pushrod from the brake pedal pushes on pistons inside the brake master cylinder. The master cylinder provides the pressure in the brake system. The force applied to the brake pedal, the ratio of brake pedal force multiplication, and the master cylinder piston size all determine how much pressure will be developed in the brake system. The size of the output pistons at the calipers and wheel cylinders, along with the brake system pressure, determines the final output force for the brakes. Use Figure 10-4 to determine brake system pressure.

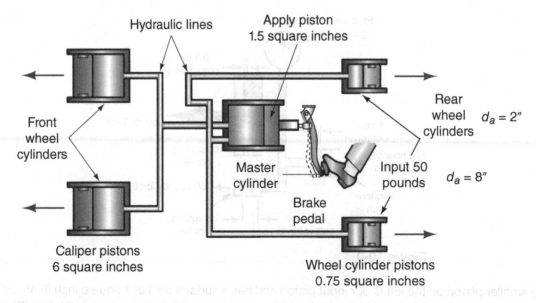

Figure 10-4

3. Input force to pedal _____

4. Brake pedal ratio D_a _____ D_b _____ Ratio _____

5. Hydraulic system pressure at master cylinder _____

6. Output force of caliper pistons _____

7. Output force of wheel cylinder pistons _____

III. Friction

Once the hydraulic system generates the pressure to actuate the pistons in the calipers and wheel cylinders, it is the job of the lining surfaces to slow the vehicle. The lining surfaces are comprised of the brake rotor or disc, brake pads, brake drum, and brake shoes. Brake pads are squeezed together against the rotor by the brake caliper. Brake fluid, under pressure from the hydraulic system, pushes on the caliper piston. As the piston moves out, pushing the inner pad against the rotor, the caliper slides inward, pressing the outer pad against the outside of the rotor. On drum brakes, the fluid pushes on two equal-diameter pistons in the wheel cylinder. The pistons move out, pressing the brake shoes against the inside surface of the brake drum.

Friction is created when uneven surfaces, in contact with each other, start to move relative to each other. Even though the two surfaces may appear smooth, there are slight imperfections that resist moving against each other. The amount of force required to move one object along another is called the coefficient of friction (CoF). Simply put, the CoF is equal to the ratio of force to move an object divided by the weight of the object, $CoF = F/M$. While several factors affect the CoF, such as temperature and speed, we will use a simple example of two bodies moving against each other. Imagine you have a 100 lb (45 kg) block of rubber on the floor of your lab. The force required to slide the block of rubber over concrete would be great. If it takes 100 lb (45 kg) of force to slide the block, the CoF will be 100/100 or 1.0. Imagine the same block of rubber now sitting on the floor of an ice hockey rink. If it only takes 15 lb (6.8 kg) of force to slide the block, what will the CoF be?

1. 15/100 = _____

To experiment with the CoF of various objects in your lab, you can make a small force gauge using an ordinary ballpoint pen, a rubber band, tape, and a paper clip. Assemble the parts as shown in Figure 10-5. Attach the paper clip to an object and try to pull it across a flat surface using the opposite end of your force meter. Measure with a ruler how much the pen tube extends from the body of the pen. This will give you an idea of how much force is being required to drag each object. Record your results below:

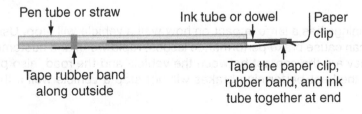

Pen tube or straw

Ink tube or dowel

Paper clip

Tape rubber band along outside

Tape the paper clip, rubber band, and ink tube together at end

Figure 10-5

2. Object _____ Surface _____ Length of extension _____

3. Object _____ Surface _____ Length of extension _____

4. Object _____ Surface _____ Length of extension _____

5. Why did some objects require more force than others did? _____
_____

6. How does the surface used to slide across affect the CoF? _____
_____

7. If a liquid is placed between the two objects, what effect will that have on the CoF? _____
_____

Static friction occurs between two stationary objects, and dynamic friction occurs between a stationary and a moving object. For example, an applied parking brake has static friction between the brake pad or shoe and the rotor or drum. When driving, there is dynamic friction between the brake pads and rotor when the driver applies the brakes.

8. List several examples of each type of friction related to the automobile.
_____
_____
_____

9. What could happen if the CoF of the brake pads or shoes was too high? _____
_____

10. What could happen if the CoF of the brake pads or shoes was too low? _____

11. What factors do you think are involved in determining the correct CoF for a particular vehicle? _____
_____
_____

The CoF of brake linings has a large impact on how well a vehicle will stop. Using lining materials with too high or low a CoF can cause brake performance issues, rapid wear, and customer dissatisfaction. Similarly, the tires, since they are the contact between the vehicle and the road, also play an important role in brake operation. Even the best-performing brakes will not stop a vehicle well if the tires cannot maintain proper traction.

IV. Brake Component Identification

1. How is the disc brake caliper in Figure 10-6a different than the caliper in Figure 10-6b?

a.

b.

Figure 10-6

2. Identify the parts of the master cylinder shown in Figure 10-7.

a. _____

b. _____

c. _____

d. _____

e. _____

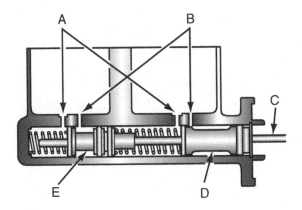

Figure 10-7

3. Identify the parts of the brake caliper shown in Figure 10-8.

a. _____

b. _____

c. _____

d. _____

e. _____

f. _____

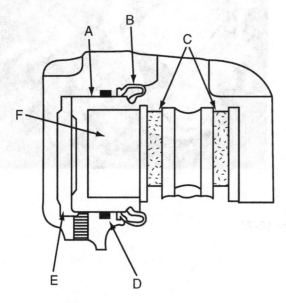

Figure 10-8

4. Identify the parts of the wheel cylinder shown in Figure 10-9.

a. _____

b. _____

c. _____

d. _____

e. _____

f. _____

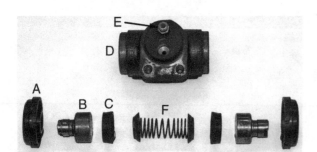

Figure 10-9

5. List examples of the different types of substances brake fluid is in contact with in the brake system.

_____

_____

_____

6. How will a leak in the hydraulic system affect brake operation? Describe how the brake pedal may feel, how stopping distances may be affected, and what the customer complaint may include. _____

_____

_____

_____

_____

Lab Activity 10-1

Name _____ Date _____ Instructor _____

Year _____ Make _____ Model _____

Brake pedal leverage

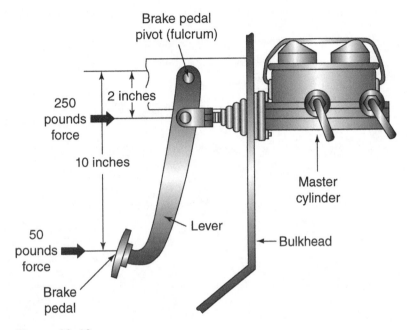

Figure 10-10

1. Using a tape measure, determine the total length of the brake pedal from the pad to the pivot. Record your measurement. _____

2. Measure the distance between the pivot and the pushrod. Record your measurement. _____

3. Determine the ratio of mechanical advantage gained by the brake pedal. _____

4. If the driver applies 100 pounds of force against the brake pedal, how much force is applied to the brake pushrod? _____

5. Describe what you think the pedal ratio may be if the vehicle were not equipped with power brakes.

   _____

   _____

6. Describe how you think the brake pedal assembly may be different on smaller and on larger vehicles than this vehicle. _____

   _____

   _____

   _____

Lab Activity 10-2

Name _____ Date _____ Instructor _____

Hydraulics

Using a selection of brake hydraulic system components provided by your instructor, determine the amount of force generated by the hydraulic system using the Figure 10-12 and the piston sizes.

Caliper piston_____   Master cylinder piston_____

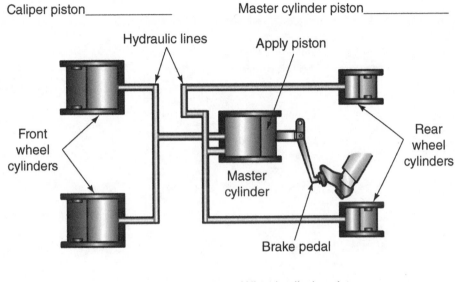

Wheel cylinder piston_____

1. Master cylinder piston diameter _____

   Master cylinder piston area _____

2. Caliper piston diameter _____

   Caliper piston area _____

3. Wheel cylinder piston diameter _____

   Wheel cylinder piston area _____

4. Using an input force of 50 pounds to the master cylinder, what is the pressure produced in the master cylinder?

   Output force of the caliper _____. Output force of the wheel cylinder _____.

5. If the master cylinder piston size is reduced, what happens to the pressure generated by the master cylinder?

   _____

   _____

Lab Activity 10-3

Name _____ Date _____ Instructor _____

Master cylinders

1. Using a selection of master cylinders provided by your instructor, measure the diameter of the primary and secondary pistons.

   Cyl 1 _____ primary   _____ secondary

   Cyl 2 _____ primary   _____ secondary

   Cyl 3 _____ primary   _____ secondary

2. Based on your measurements, determine the surface area of the pistons. Use either πr^2 or diameter$^2 \times 0.785$.

   Cyl 1 _____ primary   _____ secondary

   Cyl 2 _____ primary   _____ secondary

   Cyl 3 _____ primary   _____ secondary

3. Based on your measurements, how much pressure will be developed by each master cylinder if the input force is 200 pounds?

   Cyl 1 _____ primary psi   _____ secondary psi

   Cyl 2 _____ primary psi   _____ secondary psi

   Cyl 3 _____ primary psi   _____ secondary psi

4. Based on your measurements, how does the pressure developed by the piston vary with different-sized pistons? _____

   _____

   _____

Lab Worksheet 10-1

Name _____ Station _____ Date _____

Identify Brake Assist

ASE Education Foundation Correlation

This lab worksheet addresses the following **MLR** task:

5.A.4. Identify brake system components and configuration. **(P-1)**

Procedure

1. Year _____ Make _____ Model _____

   Determine the type of brake assist used in this vehicle.

   Vacuum boost _____   Hydraulic _____   ABS _____

2. Year _____ Make _____ Model _____

   Determine the type of brake assist used in this vehicle.

   Vacuum boost _____   Hydraulic _____   ABS _____

3. Year _____ Make _____ Model _____

   Determine the type of brake assist used in this vehicle.

   Vacuum boost _____   Hydraulic _____   ABS _____

4. Why do you think that the vehicles you inspected use that type of power brake assist?

   _____

   _____

   _____

5. Remove any tools and equipment and clean up the work area.

   Instructor's check _____

Lab Worksheet 10-2

Name _____ Station _____ Date _____

Disc ID

ASE Education Foundation Correlation

This lab worksheet addresses the following **MLR** task:

5.A.4. Identify brake system components and configuration. **(P-1)**

Description of Vehicle

Year _____ Make _____ Model _____

Engine _____ AT MT CVT PSD (circle which applies)

Procedure

1. Describe the type of front disc brake system used on this vehicle. _____
   _____

2. Caliper type: Fixed _____ Floating _____ Sliding _____

3. Number of caliper pistons _____

   Caliper piston construction material _____

4. Rotor type: Floating _____ Trapped _____

5. Describe the type of rear disc brake system used on this vehicle. _____
   _____

6. Caliper type: Fixed _____ Floating _____ Sliding _____

7. Number of caliper pistons _____

   Caliper piston construction material _____

8. Rotor type: Floating _____ Trapped _____

9. Describe the type of parking brake system used. _____
_____

10. Remove any tools and equipment and clean up the work area.

Instructor's check _____

Lab Worksheet 10-3

Name _____ Station _____ Date _____

Drum ID

ASE Education Foundation Correlation

This lab worksheet addresses the following **MLR** task:

5.A.4. Identify brake system components and configuration. **(P-1)**

Description of Vehicle

Year _____ Make _____ Model _____

Engine _____ AT MT CVT PSD (circle which applies)

Procedure

Inspect the drum brakes and determine the following:

1. Drum brake type Servo _____ Non-servo _____

2. Location of the anchor _____

3. Type of hold down springs   Coil _____   U-clip _____

                                   Spring _____   Other _____

4. Self-adjuster type   Threaded _____   Ratcheting _____

5. Describe how the self-adjuster is actuated. _____
   _____

6. How are the linings attached to the shoes?

   Riveted _____   Glued _____

7. Remove any tools and equipment and clean up the work area.

Instructor's check _____

Lab Worksheet 10-4

Name _____ Station _____ Date _____

SRS, ABS, and High-Voltage

ASE Education Foundation Correlation

This lab worksheet addresses the following **RST** task:

1.13. Demonstrate awareness of the safety aspects of supplemental restraint systems (SRS), electronic brake control systems, and hybrid vehicle high voltage circuits.

Procedure

1. Describe what dangers are present from working on or near SRS/SIR systems. _____
   _____
   _____

2. What colors are used to identify SRS/SIR system wiring and components? _____

3. Explain why antilock brake systems can pose a danger when working on the brakes. _____
   _____
   _____

4. What color is used to identify the high-voltage wiring on hybrid and electric vehicles? _____.
   _____

5. What precautions should be taken when working on a hybrid or electric vehicle? _____
   _____
   _____

Instructor's check _____

CHAPTER 11

Brake System Service

Review Questions

1. Line or _____ nut wrenches are used on brake line fittings.

2. A _____ tool is used to make new flares on a brake line.

3. A _____ bleeder applies pressure to the brake fluid at the master cylinder and forces air and fluid out of the hydraulic system.

4. Do not use _____-end wrenches to break loose tight fasteners.

5. Do not allow brake fluid to remain in contact with _____ surfaces, _____, plastics, or any other parts of the vehicle.

6. As a technician, you will be required to _____-_____ vehicles to verify customer concerns and to verify repairs have been completed properly.

7. Describe four problems you are checking for during a test drive.

   a. _____

   b. _____

   c. _____

   d. _____

8. The brake pedal assembly may be inspected for pedal _____, travel, and _____ play.

9. Incorrect brake pedal height may be caused by which of the following concerns?

   a. Low brake fluid level

   b. Improper parking brake adjustment

   c. Excessive shoe-to-drum clearance

   d. Faulty proportioning valve

10. Brake pedal _____ is the very slight movement of the pedal before the brake pushrod begins to move.

11. List systems that may use the brake light switch as an input. _____
_____
_____

12. Explain how to test brake light switch operation. _____
_____
_____

13. Referring to Figure 11-1: *Technician A* says the adjustment shown is used to adjust the stoplight switch. *Technician B* says the adjustment shown is to preset the pushrod into the power brake booster. Who is correct?

   a. Technician A

   b. Technician B

   c. Both A and B

   d. Neither A nor B

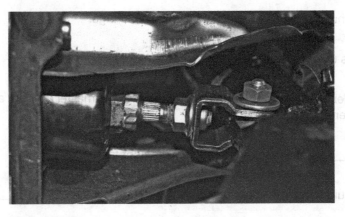

Figure 11-1

14. Explain why the brake light circuit is often a hot-at-all-times circuit. _____
_____
_____

15. To test for a concern with power-adjustable pedals, a _____ is needed to access DTCs and data.

16. When it is necessary to add brake fluid to the reservoir, always use _____ brake fluid from a _____ container.

17. Describe two causes for the brake fluid level to be low._____
_____
_____
_____

18. Explain why many vehicle manufacturers recommend periodic brake fluid changes. _____
_____
_____
_____

19. Which of the following does not cause the red brake warning light to illuminate?

    a. Low brake fluid level

    b. Contaminated brake fluid

    c. Loss of hydraulic pressure

    d. Activated parking brake

20. A late model vehicle with about 35,000 miles has low brake fluid level and the red BRAKE light is illuminated on the dash. Explain what can cause this and what should be recommended to the customer.

    _____

    _____

    _____

    _____

21. The _____ - _____ switch will turn on the red BRAKE light if there is an imbalance in pressure in the brake system.

22. Explain what can result if a petroleum-based fluid is added to the brake fluid.

    _____

    _____

    _____

23. Explain what happens to brake fluid when it has been left in service for too long.

    _____

    _____

    _____

    _____

24. A faulty flexible brake hose can cause which of the following brake concerns?

    a. Sticking brake caliper

    b. Non-functioning brake caliper

    c. Fluid loss

    d. All of the above

25. Over time, moisture absorbed by the brake fluid can cause the wheel cylinder pistons to _____ into their bores.

26. External fluid leaks from the master cylinder result from a leaking seal on the _____ piston.

27. *Technician A* says a leaking master cylinder primary piston seal will always result in fluid showing on the outside of the power brake booster. *Technician B* says some master cylinders can leak into the power brake booster assembly. Who is correct?

    a. Technician A

    b. Technician B

    c. Both A and B

    d. Neither A nor B

28. Describe the steps to replace a master cylinder. _____

_____

_____

_____

29. Explain the steps to replace a leaking brake line.

_____

_____

_____

30. Label the type of brake line flare Figure 11-2.

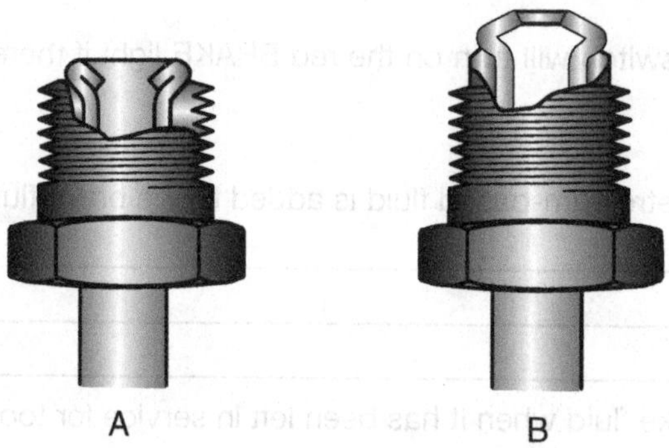

A   B

Figure 11-2

31. A vehicle's brake pedal sinks to the floor under normal pressure. Upon inspection, fluid level is correct and no external leaks are discovered. Which of the following is the most likely cause?

    a. Worn brake pads and rotors

    b. Kinked brake line

    c. Bypassing master cylinder

    d. None of the above

32. Which of the following is not a benefit of periodic brake fluid flushing?

    a. Decreased rust formation in hydraulic system

    b. Prolong service life of hydraulic components

    c. Reduced brake fluid boiling temperature

    d. Reduced risk of brake fade

33. Explain in detail three methods of brake system bleeding. _____

_____

_____

_____

_____

34. To properly bleed the ABS, a _____ may be required.

35. *Technician A* says the tool shown in Figure 11-3 is required to bleed ABS-equipped vehicles. *Technician B* says the tool shown in Figure 11-3 is used to bleed the master cylinder. Who is correct?

    a. Technician A

    b. Technician B

    c. Both A and B

    d. Neither A nor B

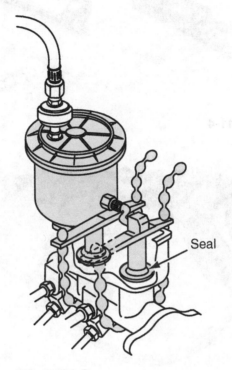

Figure 11-3

Activities

1. Label the brake tools shown in Figure 11-4 and Figure 11-5.

Figure 11-4

Figure 11-5

2. Describe the purpose and operation of the pressure differential valve. _____
    _____
    _____
    _____

3. Describe the purpose and operation of the proportioning valve. _____
    _____
    _____
    _____

4. Describe the purpose and operation of the load-sensing proportioning valve.
    _____
    _____
    _____

5. Explain the purpose and operation of the component shown in Figure 11-6. _____

_____

_____

_____

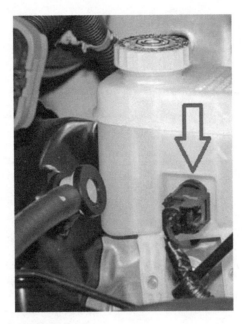

Figure 11-6

5. Explain the purpose and operation of the component shown in Figure 11-6.

_____

_____

_____

Figure 11-6

Lab Activity 11-1

Name _____ Date _____ Instructor _____

Year _____ Make _____ Model _____

Brake pedal inspection

1. Locate and record the following brake pedal specs.

   Pedal free play _____

   Pedal height _____

   Pedal travel _____

   a. Why is brake pedal free play important for proper brake system operation?

   _____

   _____

   b. What other components can affect brake pedal height? _____

   _____

2. Operate the brake pedal and listen for noise. Note your findings. _____

   _____

3. Check and record the brake pedal free play. _____

   a. Is the free play within specs?   Yes _____   No _____

      If no, how can free play be adjusted? _____

4. Check and record pedal height. _____

   a. Is pedal height within specs?   Yes _____   No _____

      If no, what affects pedal height? _____

   _____

5. Check and record pedal travel. _____

   a. Is pedal travel within specs?   Yes _____   No _____

      If no, what affects pedal travel? _____

   _____

6. Based on your inspections, what actions are required? _____

   _____

Lab Activity 11-2

Name _____ Date _____ Instructor _____

Year _____ Make _____ Model _____

Brake light inspection

1. Locate the brake light switch and note its location. _____

2. Have an assistant press the brake pedal and note the operation of the brake lights.
   Are all lights operating?   Yes _____   No _____
   If no, note which lights do not work? _____
   If none of the brake lights operate what should be checked?
   _____

3. Locate the specification for the amount of brake pedal travel required to activate the brake light circuit.
   _____

4. Measure the brake pedal travel necessary to activate the brake lights and note your results.
   _____

5. Is the brake light switch properly adjusted?   Yes _____   No _____

6. Explain the procedure to adjust the brake light switch. _____
   _____
   _____

Lab Worksheet 11-1

Name _____ Station _____ Date _____

Describe Test Drive

ASE Education Foundation Correlation

This lab worksheet addresses the following **MLR** task:

5.A.2. Describe procedure for performing a road test to check brake system operation, including an anti-lock brake system (ABS). **(P-1)**

Description of Vehicle

Year _____ Make _____ Model _____

Engine _____ AT MT CVT PSD (circle which applies)

Procedure

1. Before performing a road test of a vehicle, what are some specific items you should check to make sure the vehicle is safe to drive? _____
   _____
   _____

   Taking a vehicle on a road test usually requires you to drive the vehicle on public roads. Always follow all local traffic laws and operate the vehicle in a safe and professional manner.

2. What type of driving and road conditions should the vehicle be driven on to check the brake system operation? _____
   _____

3. Before driving the vehicle: Turn the ignition on and note the ABS light; does the light illuminate?

   Yes _____   No _____

   If no, what can this indicate? _____

   If yes, what does this indicate? _____

4. When performing a test drive, note if any of these conditions are present. Check all that apply.

Noise when braking _____   Pulls when braking _____

Pulsation in brake pedal _____   Pulsation in steering wheel _____

Brakes grab _____   Brakes drag _____

ABS activates _____   ABS light comes on _____

5. Based on your road test, what are possible causes of the concerns noted while driving?_____

_____

_____

Lab Worksheet 11-2

Name _____ Station _____ Date _____

Describe Pedal Feel

ASE Education Foundation Correlation

This lab worksheet addresses the following **MLR** task:

5.B.1. Describe proper brake pedal height, travel, and feel. **(P-1)**

Description of Vehicle

Year _____ Make _____ Model _____

Engine _____ AT MT CVT PSD (circle which applies)

Procedure

1. With the engine off, describe how the brake pedal should feel when pressed. _____

   _____

   _____

2. With the engine running, describe how the brake pedal should feel and react when pressed. _____

   _____

   _____

3. If the pedal sinks to the floor, what does this indicate? _____

   _____

4. If the pedal is low, soft, or does not feel firm when pressed, what can this indicate? _____

   _____

   _____

Lab Worksheet 11-3

Name _____ Station _____ Date _____

Inspect Master Cylinder

ASE Education Foundation Correlation

This lab worksheet addresses the following **MLR** task:

5.B.2. Check master cylinder for external leaks and proper operation. **(P-1)**

Description of Vehicle

Year _____ Make _____ Model _____

Engine _____ AT MT CVT PSD (circle which applies)

Procedure

1. Locate the master cylinder and note the following:

Number of brake lines attached _____

Location of reservoir _____

Fluid level sensor   Yes _____   No _____

2. With the engine off, pump the brake pedal 10 times and hold.

Does the pedal slowly sink to the floor?   Yes _____   No _____

If yes, what can cause this? _____

_____

3. While keeping pressure on the brake pedal, start the engine.

Does the pedal drop slightly?   Yes _____   No _____

If yes, what does this indicate? _____

Does the pedal sink to the floor?   Yes _____   No _____

If yes, what does this indicate? _____

4. Inspect the area where the master cylinder is attached to the brake booster.

Is brake fluid or any wetness present?   Yes _____   No _____

If yes, what does this indicate? _____

5. Inspect the brake line connections at the master cylinder.

   Is brake fluid or any wetness present?   Yes _____   No _____

   If yes, what does this indicate? _____

6. Based on your inspection, what is the condition of the master cylinder? _____

   _____

   _____

7. Remove any tools and equipment and clean up the work area.

                                        Instructor's Check _____

Lab Worksheet 11-4

Name _____ Station _____ Date _____

Inspect Brake Lines and Hoses

ASE Education Foundation Correlation

This lab worksheet addresses the following **MLR** task:

5.B.3. Inspect brake lines, flexible hoses, and fittings for leaks, dents, kinks, rust, cracks, bulging, wear, and loose fittings/supports. **(P-1)**

Description of Vehicle

Year _____ Make _____ Model _____

Engine _____ AT MT CVT PSD (circle which applies)

Procedure

Safely raise and support the vehicle.

1. Install fender covers and inspect the master cylinder and line. Check for signs of leakage at the rear of the master cylinder, at the fittings attached to the master cylinder, and from the lines. Note your findings.

   _____

   _____

2. Check the following for rust through, kinks, dents, bulges, and other forms of damage.

Lines from the master cylinder to a valve or ABS unit	OK _____	Not OK _____
Lines from the valve or ABS unit to each wheel	OK _____	Not OK _____
Connections from steel lines to rubber hoses	OK _____	Not OK _____
Rubber hoses	OK _____	Not OK _____
Connections of hoses to calipers	OK _____	Not OK _____
Line clamps and hose brackets	OK _____	Not OK _____

3. Based on your inspection, what is the necessary action for this vehicle? _____

   _____

4. Once finished, put away all tools and equipment and clean up the work area.

   Instructor's Check _____

Lab Worksheet 11-5

Name _____ Station _____ Date _____

ID Brake Warning Light Components

ASE Education Foundation Correlation

This lab worksheet addresses the following **MLR** task:

5.B.5. Identify components of the hydraulic brake warning light system. **(P-3)**

Description of Vehicle

Year _____ Make _____ Model _____

Engine _____ AT MT CVT PSD (circle which applies)

Procedure

1. Turn the ignition on and note the brake warning lamp.   On _____   Off _____

   If the lamp does not illuminate, what might this mean? _____

   _____

   What steps would you take to diagnose this problem? _____

   _____

2. Start the engine and note the warning lamp.   Remains on _____   Off _____

   List three possible causes for the lamp remaining on with the engine running.

   _____

   _____

   _____

3. If the lamp is off, carefully apply and release the parking brake to check warning light operation. The light should illuminate with the parking brake set and turn off when released.

   Light on _____   Light remains off _____

   If the light remains off, what would you need to check? _____

   _____

4. Is the vehicle is equipped with a brake fluid level sensor?   Yes _____   No _____

5. If the fluid sensor is built into the master cylinder cap, remove the cap and note if the brake warning lamp illuminates.   Yes _____   No _____

6. If the sensor is built into the reservoir, unplug the sensor and jump the two terminals together and note if the warning lamp illuminates.   Yes _____   No _____

7. If the light does not illuminate, what should be checked? _____
_____

8. Remove any tools and equipment and clean up the work area.
                                              Instructor's Check _____

Lab Worksheet 11-6

Name _____ Station _____ Date _____

Bleed System

ASE Education Foundation Correlation

This lab worksheet addresses the following **MLR** task:

5.B.6. Bleed and/or flush brake system. **(P-1)**

Description of Vehicle

Year _____ Make _____ Model _____

Engine _____ AT MT CVT PSD (circle which applies)

Procedure

1. Before attempting to bleed the brake system, first locate and record the bleeding procedures outlined in the manufacturer's service information. _____

   _____

   _____

2. Locate the bleeder valves and make sure each will open. Do not force the bleeder valves as they break easily. Note any problems opening the bleeder valves. _____

   _____

3. If any or all bleeder valves are seized, what actions may be required to bleed the brake system?

   _____

   _____

4. Which type of bleeding procedure is to be used?

   Manual bleeding _____   Pressure bleeding _____

   Vacuum bleeding _____   Other _____

   Is a scan tool required?   Yes _____   No _____

5. Bleed the brake system following the manufacturer's procedures until clean fluid comes out of each bleeder valve. Note any problems you encounter.

_____

_____

Instructor's Check _____

6. Remove any tools and equipment and clean up the work area.

Instructor's Check _____

Lab Worksheet 11-7

Name _____ Station _____ Date _____

Test Brake Fluid

ASE Education Foundation Correlation

This lab worksheet addresses the following **MLR** task:

5.B.7. Test brake fluid for contamination. **(P-1)**

Description of Vehicle

Year _____ Make _____ Model _____

Engine _____ AT MT CVT PSD (circle which applies)

Procedure

1. Install fender covers and inspect the brake fluid level and condition. Note your findings.

   _____

   _____

2. Describe how brake fluid can become contaminated._____

   _____

3. Remove the master cylinder cap and use a brake fluid moisture tester to test the brake fluid. Note your findings.

   Brake fluid boiling point _____

4. Test the brake fluid with a brake fluid test strip. Note the results of the test._____

   _____

5. Reinstall the master cylinder cap.

6. Based on your inspection, what are the necessary actions for this vehicle?_____

   _____

7. Once finished, put away all tools and equipment and clean up the work area.

                          Instructor's Check _____

Lab Worksheet 11-8

Name _____ Station _____ Date _____

Check Parking Brake Operation

ASE Education Foundation Correlation

This lab worksheet addresses the following **MLR** task:

5.F.3. Check parking brake operation and parking brake indicator light system operation; determine necessary action. **(P-1)**

Description of Vehicle

Year _____ Make _____ Model _____

Engine _____ AT MT CVT PSD (circle which applies)

Procedure

1. Describe the type and operation of the parking brake system used on this vehicle.

   _____

   _____

2. Apply the parking and note the feel of the parking brake handle or pedal.
   Is the parking brake very loose?   Yes _____   No _____
   If yes, what may this indicate? _____
   Is the parking brake very tight?   Yes _____   No _____
   If yes, what may this indicate? _____

3. Apply the parking brake and note if the BRAKE warning light illuminates on the dash with the key in the ON positon.
   Did the light turn on?   Yes _____   No _____
   If no, what may this indicate? _____

   _____

   Does the light go out when the brake is released?   Yes _____   No _____

4. Based on your inspection, what is the condition of the parking brake system?_____

   _____

   _____

Instructor's Check _____

Lab Worksheet 11-9

Name _____ Station _____ Date _____

Check Stop Lights

ASE Education Foundation Correlation

This lab worksheet addresses the following **MLR** task:

5.F.4. Check operation of brake stop light system. **(P-1)**

Description of Vehicle

Year _____ Make _____ Model _____

Engine _____ AT MT CVT PSD (circle which applies)

Procedure

1. Locate the brake light switch and note its location. _____

2. Have an assistant press the brake pedal and note the operation of the brake lights.

    Are all lights operating?   Yes _____   No _____

    If no, note which lights do not work? _____

    If none of the brake lights operate what should be checked? _____

    _____

3. Locate the specification for brake pedal travel required to activate the brake light circuit.

    Specification _____

4. Measure the brake pedal travel necessary to activate the brake lights and note your results._____

    _____

5. Is the brake light switch properly adjusted?   Yes _____   No _____

6. Explain the procedure to adjust the brake light switch. _____

_____

_____

7. Remove any tools and equipment and clean up the work area.

Instructor's Check _____

Lab Worksheet 11-10

Name _____ Station _____ Date _____

Select and Handle Brake Fluid

ASE Education Foundation Correlation

This lab worksheet addresses the following **MLR** task:

5.B.4. Select, handle, store, and fill brake fluids to proper level; use proper fluid type per manufacturer specification. **(P-1)**

Description of Vehicle

Year _____ Make _____ Model _____

Engine _____ AT MT CVT PSD (circle which applies)

Procedure

1. Install fender covers. Inspect the cap of the master cylinder reservoir. Clean the top of the cap if necessary. Note what type of brake fluid is recommended for this vehicle.

   Brake fluid type _____

2. If the fluid type is not listed on the reservoir cap, refer to the manufacturer's service information to determine the correct type of fluid.

3. Many master cylinder reservoirs are opaque enough to see the fluid level without removing the cap. If this is possible, inspect and note the brake fluid level._____

   _____

4. If the fluid level cannot be seen, clean the reservoir cap and remove the cap to check the fluid level. Note the condition of the seal under the cap and the fluid level. _____

5. If the fluid level is low, add enough fluid to bring the level up the MAX line. Use only new fluid from a sealed container.

   Instructor's check _____

6. Describe the proper way to store brake fluid. _____

_____

_____

7. Why is it important to keep brake fluid sealed once it has been opened?

_____

_____

8. What are several problems that can be caused by using brake fluid that has had excessive exposure to air?

_____

_____

9. Once finished, put away all tools and equipment and clean up the work area.

Instructor's check _____

CHAPTER 12

Drum Brake System Principles

Review Questions

1. Drum brakes use a set of brake _____ that expand outward against the inside of the rotating brake _____.

2. _____ pressure acts on the pistons in the wheel cylinders to press the shoes outward.

3. Explain the advantage that drum brakes utilize to increase the amount of stopping power available compared to disc brakes. _____
   _____
   _____

4. List four disadvantages of drum brakes compared to disc brakes. _____
   _____
   _____
   _____

5. The _____ holds the components of the drum brake assembly.

6. Brake _____ are curved pieces of metal onto which the friction lining is applied.

7. True or False: All four brake shoes are identical and can be installed on either axle and in either the front or the rear position _____.

8. Compare and contrast a brake drum and a brake rotor. _____
   _____
   _____

9. The shoes are pulled back from the drum by the _____ springs.

10. Holddown springs and _____ hold the shoes to the backing plate.

11. Hydraulic pressure forces the two pistons inside of the _____ outward and against the shoes.

12. Drum brakes use a _____ - _____ to maintain the shoe-to-drum clearance as the brakes wear.

13. The location of the _____ determines whether the brakes are servo or nonservo.

14. Brakes that use leverage from one shoe against the other shoe to increase brake application force are called:

    a. Duo-servo brakes

    b. Self-energizing brakes

    c. Dual-servo brakes

    d. Duo-servo brakes, self-energizing brakes, and dual-servo brakes

15. Some drum brake designs use the self-adjuster as part of the _____ brake.

16. Nonservo brakes are also called _____ - _____ brakes.

17. The most common types of self-adjusters are _____ on one end and freely rotate on the other.

18. A _____ type self-adjuster is operated by using the parking brake _____.

19. The parking brake operation on a vehicle with rear drum brakes is partially dependent on how well the _____ are adjusted.

20. True or False: Only vehicles with manual transmissions really need to use the parking brake.

21. Parking brakes can be activated by:

    a. A hand-operated lever

    b. An electrical input

    c. A foot-operated pedal

    d. A hand-operated level, an electrical input, and a foot-operated pedal

22. Describe the two types of parking brake cables. _____

    _____

    _____

    _____

23. *Technician A* says all parking brake systems are mechanically operated. *Technician B* says some vehicles use electrically operated parking brakes. Who is correct?

    a. Technician A

    b. Technician B

    c. Both A and B

    d. Neither A nor B

Activities

1. Identify the components of the drum brakes shown in Figure 12-1.

   A. _____   B. _____

   C. _____   D. _____

   E. _____   F. _____

   G. _____   H. _____

   I. _____   J. _____

   K. _____

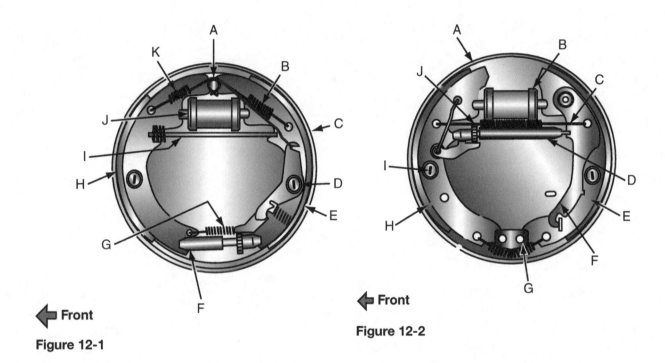

◀ **Front**

Figure 12-1

◀ **Front**

Figure 12-2

2. Identify the components of the drum brake shown in Figure 12-2.

   A. _____   B. _____

   C. _____   D. _____

   E. _____   F. _____

   G. _____   H. _____

   I. _____   J. _____

3. Identify the components of the wheel cylinder shown in Figure 12-3.

A. _____   B. _____

C. _____   D. _____

E. _____   F. _____

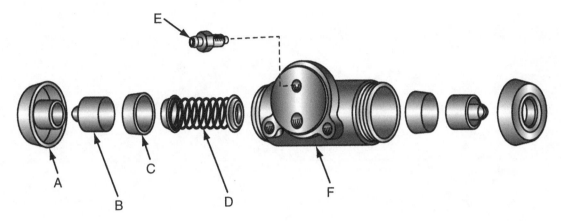

Figure 12-3

3. Identify the components of the wheel cylinder shown in Figure 12-3.

A. _____ B. _____
C. _____ D. _____
E. _____ F. _____

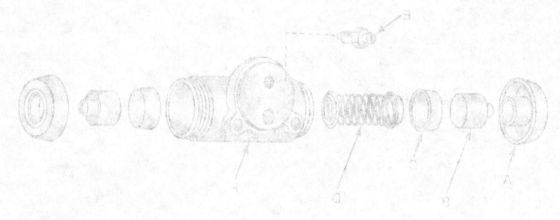

Figure 12-3

Lab Activity 12-1

Name _____ Date _____ Instructor _____

Year _____ Make _____ Model _____

Identify Brake Type and Components

Inspect the drum brakes and determine the following:

1. Drum brake type Servo _____ Nonservo _____

2. Location of the anchor _____

3. Type of holddown springs   Coil _____   U-clip _____

    Spring _____   Other _____

4. Self-adjuster type   Threaded _____   Ratcheting _____

5. Describe how the self-adjuster is actuated. _____
   _____

6. How are the linings attached to the shoes?

    Riveted _____   Glued _____

7. Locate and record the minimum lining thickness. _____

8. Locate and record the maximum drum diameter. _____

CHAPTER 13

Drum Brake System Inspection and Service

Review Questions

1. Explain why it is important to use the proper brake tools and equipment. _____
   _____
   _____

2. A vacuum enclosure or a _____ is used to clean brake dust from the brake assembly.

3. A vacuum or sink is used to trap airborne dust and _____ fibers that may be present in the linings.

4. A _____ spring tool is used to safely remove and install high-tension _____ springs.

5. To measure brake drum wear and out-of-round, a drum brake _____ is used.

6. Do not use _____ or _____ in place of the proper brake tools.

7. Before beginning to disassemble a drum brake assembly, locate a diagram or take a _____ of the components to help with reassembly.

8. List five service precautions you should follow before you begin to service the brakes.
   a. _____
   b. _____
   c. _____
   d. _____
   e. _____

9. Describe how to safely contain asbestos dust when working on brake systems.

   _____

   _____

   _____

   _____

10. Any time a vehicle is checked for a brake system complaint, _____ wheel brake assemblies should be inspected.

11. A full brake inspection, including the operation of the _____ brake and warning lights, should be performed any time a brake concern is present.

12. All of the following can cause drum brake noise except:

    a. Excessive dust trapped in the drum

    b. High metal content in the linings

    c. Shoe wear indicator

    d. Inadequate lubrication between the shoes and backing plate

13. _____ is where the brake applies too quickly or with too much force, which causes the wheel to lock.

14. *Technician A* says an out-of-round drum may make the whole car shake when the brakes are applied. *Technician B* says a brake pulsation that disappears when only the parking brake is applied that is caused by the front disc brakes. Who is correct?

    a. Technician A

    b. Technician B

    c. Both A and B

    d. Neither A nor B

15. A _____ brake pedal is usually caused by excessive shoe-to-drum clearance or _____ shoes and drums.

16. Before you attempt to remove the brake drum, check the _____ for the procedure to remove the drum.

17. *Technician A* says all modern cars and trucks used floating drums. *Technician B* says floating drums require the rear axle to be removed to remove the drum. Who is correct?

    a. Technician A

    b. Technician B

    c. Both A and B

    d. Neither A nor B

18. Regarding Figure 13-1: *Technician* A says the retainer should be reinstalled after the brakes are inspected. *Technician B* says the retainer can be discarded once the drum is removed. Who is correct?

    a. Technician A

    b. Technician B

    c. Both A and B

    d. Neither A nor B

Figure 13-1

19. Describe how to remove a floating brake drum. _____
    _____
    _____

20. The three most common nonfloating drum arrangements are the _____ bearing, _____ wheel bearing, and _____ -floating rear axle types.

21. A common method of removing full-floating drums involves unbolting the _____ flange from the hub and removing the axle.

22. Two technicians are discussing the brake dust shown in Figure 13-2. *Technician A* says the dust is wet from a leaking wheel cylinder. *Technician B* says the wet dust can be cleaned off and the shoes remain in service. Who is correct?

    a. Technician A

    b. Technician B

    c. Both A and B

    d. Neither A nor B

Figure 13-2

23. What is shown in Figure 13-3?
    a. Measuring brake shoe width
    b. Measuring lining thickness
    c. Adjusting the parking brake
    d. Checking shoe-to-drum clearance

Figure 13-3

24. Hard spots in the drum are caused by _____ the brakes, causing the metal to change under heat stress.

25. _____ in the drum facing, such as around lug holes, can occur from extreme stress or from a collision.

26. A _____ -mouthed drum is one in which the inside diameter is less than the diameter around the outside of the drum.

27. Drum brake _____ consists of the springs and related parts of the drum brake assembly.

28. What inspection is being performed in Figure 13-4?

    a. Wheel cylinder leak

    b. Axle seal leak

    c. Shoe lining thickness

    d. None of the above

Figure 13-4

29. Describe how to check parking brake operation. _____

    _____

    _____

    _____

30. Begin drum brake disassembly by cleaning as much brake _____ from the linings and hardware as possible with either a wet sink or brake dust vacuum.

31. If you are working on brakes that you are not familiar with, take a _____ of the brake before you begin disassembly.

32. Describe how and why should the backing plate be inspected? _____

    _____

    _____

    _____

33. Bonded linings should be replaced when the lining wears to a thickness of _____ inch or about _____ mm.

34. Many technicians recommend replacing the _____ and _____ when the shoes are replaced.

35. The brake drum should be replaced if it has which of the following defects?

    a. Cracks in the friction surface

    b. Exceeds maximum diameter

    c. Deep scoring

    d. All of the above

36. Before attempting to refinish a brake drum, first measure the drum diameter with a drum brake

    _____.

37. When refinishing a drum, what is the final step before reinstalling the drum on the vehicle?

    a. Applying a nondirectional finish to the drum surface

    b. Washing the drum thoroughly inside to remove dust

    c. Lubricate the drum surface with high-temperature brake lubricant

    d. Reapplying drum balance weights removed during refinishing

38. List the steps in replacing a wheel cylinder. _____

    _____

    _____

    _____

39. Which of the following lubricants may be used when reassembling the drum brakes?

    a. Chassis grease

    b. WD-40

    c. Silicone spray

    d. None of the above

40. What action is shown in Figure 13-5?

    a. Measuring the brake drum diameter

    b. Determining shoe lining thickness

    c. Measuring to preadjust the shoes

    d. Measuring hub runout

Figure 13-5

41. A vehicle with a disc and drum brake system has a low service brake and the parking brake does not hold when applied. *Technician A* says the parking brake adjustment is incorrect. *Technician B* says the drum brake self-adjusters may not be working. Who is correct?

    a. Technician A

    b. Technician B

    c. Both A and B

    d. Neither A nor B

Activities

1. Match the drum brake concern with the most likely cause.

Brake pulsation	Sticking parking brake cable
Brake grab	Bent backing plate
Brake noise	Fluid contamination
Abnormal lining wear	Excessive dust buildup
Brake drag	Out-of-round drum

Lab Activity 13-1

BRAKE SYSTEM INSPECTION

Date _____

Name _____

Address _____

City _____ State _____ Zip Code _____

Year _____ Make _____ Model _____

Mileage _____ License _____ VIN _____ Eng _____

Customer Interview:

_____

_____

_____

_____

_____

BRAKE INSPECTION CHECKLIST

Owner _____ Phone _____ Date _____

Address _____ VIN _____

Make _____ Model _____ Year _____ Lic. number _____ Mileage _____

Road test results		Yes	No	Right	Left
Brake pulsation	Above 30 MPH				
	Below 30 MPH				
	Bump steer				
Steering wheel movement					
Pull under braking					
Noise under braking					
ABS activation					

	OK	Not ok
Power assist operation		

	Yes	No
ABS light on/off		

	OK	Not ok		
Parking brake travel	OK	Not ok		
Parking brake cables	OK	Not ok		

Underhood checks	OK	Not ok
Brake fluid level		
Brake fluid moisture content		
Brake fluid copper content		
Master cylinder reservoir cover/gasket		
Master cylinder		
Lines and fittings		

Power booster	OK	
Vacuum supply		
Vacuum hose and check valve		

	OK	Not ok
Electric parking brake operation	OK	Not ok

comments

Rotor thickness	Specs Frnt _____ Rear _____
measured	

RF _____ LF _____ RR _____ LR _____

Rotor runout	Specs Frnt _____ Rear _____
measured	

RF _____ LF _____ RR _____ LR _____

Rotor parallelisms	Specs Frnt _____ Rear _____
measured	

RF _____ LF _____ RR _____ LR _____

Drum diameter	Specs Frnt _____ Rear _____
measured	

RF _____ LF _____ RR _____ LR _____

Brake hoses and lines	OK	Not
Caliper fluid leaks		
Hardware		
Mounting bolts		
Bracket bolts		
Guides and pins		
Boots		
Clips, springs, other		

Drum condition	OK	Not
Brake hoses and lines		
Wheel cylinder fluid leaks		
Hardware		
Hold down springs		
Return springs		
Self adjuster		
Parking brake		
Clips		

Lining thickness	Spec Fnt _____ Rear _____
measured	

RF _____ LF _____ RR _____ LR _____

Lining thickness	Spec Fnt _____ Rear _____
Comments	

Others	OK	Not OK
Wheel bearings		
Control arm bushings		
Strut rod bushings		
Radius arm bushings		
Others		

Lab Worksheet 13-1

Name _____ Station _____ Date _____

Inspect Drum

ASE Education Foundation Correlation

This lab worksheet addresses the following **MLR** task:

5.C.1. Remove, clean, and inspect brake drum; measure brake drum diameter; determine serviceability. **(P-1)**

Description of Vehicle

Year _____ Make _____ Model _____

Engine _____ AT MT CVT PSD (circle which applies)

Procedure

1. Inspect the outside of the brake drum for damage. Note your findings. _____
   _____
   _____

2. With the drum still on the vehicle, spin the drum by hand, and note the following:

   Spins easily   Yes _____   No _____

   Continues to spin several times once spun   Yes _____   No _____

   Gets tight and loose as it spins   Yes _____   No _____

   Noise when spinning   Yes _____   No _____

3. Remove the drum and inspect the friction surface for scoring, cracks, pitting, and bluing. Note your findings.
   _____
   _____

4. Locate the drum diameter specifications and record.

    Max diameter _____

    Service limit or machine to specification _____

5. Using a drum micrometer, measure the drum and record your readings.

    Diameter _____

    Out-of-round _____

6. Examine the drum for signs of bell-mouthing, concave or convex wear, heat cracks, or checking. Note your findings. _____
    _____
    _____

7. Based on your inspection, what are your conclusions about the drum? _____
    _____
    _____

8. Remove any tools and equipment and clean up the work area.

    Instructor's check _____

Lab Worksheet 13-2

Name _____ Station _____ Date _____

Refinish Drum

ASE Education Foundation Correlation

This lab worksheet addresses the following **MLR** task:

5.C.2. Refinish brake drum and measure final drum diameter; compare with specifications. **(P-1)**

Description of Vehicle

Year _____ Make _____ Model _____

Engine _____ AT MT CVT PSD (circle which applies)

Procedure

1. Locate the brake drum wear specifications.

   Machine to specification _____

   Discard specification _____

2. Measure the brake drum, vertically and horizontally to check for out-of-round.

   Measurement _____

   Measurement _____

3. If the drum can be refinished, select the correct adaptors and mount the drum to the brake lathe. Install the vibration damper and adjust the latch as needed to perform a scratch cut.

   Instructor's check _____

4. Perform a scratch cut on the drum's surface. Next, loosen the drum on the lathe and rotate the drum 180 degrees from its original mounting and tighten. Perform a second scratch cut and compare the two cuts. What does the scratch cut indicate? _____

   _____

   _____

5. If the drum and lathe setup is correct, set the lathe to perform a fast cut across the drums surface and engage the feed.

Instructor's check _____

6. Once the fast cut is complete, turn the lathe off and inspect the drum's surface. If the out-of-round is corrected and the surface is free of defects, perform a slow final cut. If the out-of-round or surface defects still exist, perform a second fast cut.

Instructor's check _____

7. Once the slow cut is finished, turn the lathe off and remove the drum. Clean the friction surface with warm soapy water to remove the fine dust present from the refinishing process.

Instructor's check _____

8. What can result from not cleaning the drum after refinishing? _____

_____

_____

9. Measure the drum with the micrometer to ensure it is not beyond the machine to or discard specifications. Measurement. _____

10. If the drum diameter is greater than specifications, what is the necessary action? _____

11. Once finished, put away all tools and equipment and clean up the work area.

Instructor's check _____

Lab Worksheet 13-3

Name _____ Station _____ Date _____

Service Brake Shoes

ASE Education Foundation Correlation

This lab worksheet addresses the following **MLR** tasks:

5.C.3. Remove, clean, inspect, and/or replace brake shoes, springs, pins, clips, levers, adjusters/self-adjusters, other related brake hardware, and backing support plates; lubricate and reassemble. **(P-1)**

5.C.5. Pre-adjust brake shoes and parking brake; install brake drums or drum/hub assemblies and wheel bearings; make final checks and adjustments. **(P-1)**

Description of Vehicle

Year _____ Make _____ Model _____

Engine _____ Rear brake type _____

Procedure

Safely raise and support the vehicle.

1. Refer to the manufacturer's service information for inspection and service procedures for the drum brake system. Summarize these procedures. _____

   _____

   _____

2. Remove the brake drum. This may require removing any retaining clips used to hold the drum in place during vehicle assembly. Remove and discard the clips if applicable. If the shoes are adjusted out, drum removal can be difficult as a ridge is formed at the outside edge of the drum's friction surface. Backing off the shoe adjuster may be required for drum removal. Describe any problems you may have had in removing the drum.

   _____

   _____

3. Inspect the drum brake assembly for signs of fluid contamination from a leaking wheel cylinder or axle seal. Note your findings. _____

   _____

4. Using an OSHA approved method, remove the brake dust from the drum and brake assembly.

WARNING—Do not blow the brake dust off with compressed air. Brake dust can contain asbestos and other harmful chemicals. Always remove brake dust using an OSHA approved dust collection vacuum or wet sink.

Instructor's check _____

5. Before beginning to disassemble the brakes, take a picture or draw a picture of how the brakes are installed, where springs attach, and the location of hardware. This will help with reassembly. Begin disassembly and place each component on a workbench arranged as it was installed.

Instructor's check _____

6. Once the brake shoes and hardware have been removed, inspect the backing plate for signs of damage and wear. Refer to the service information for inspection guidelines. Note your findings.

_____

_____

7. Clean the backing plate of any remaining dust using the OSHA approved brake dust containment equipment. Locate the raised pads on the backing plate and clean with a wire brush if necessary. Apply a light coat of lubricant to the raised pads and any other locations as indicated in the service information.

Lubricant used _____

Instructor's check _____

8. Inspect the wheel cylinder for fluid leaks. Pull the dust boot away from the cylinder and note any fluid inside the boot. Note your findings. _____

_____

9. Inspect the brake hardware for excessive rust, damage, and wear. Disassemble the self-adjuster and inspect it for wear and damage. Note your findings. _____

_____

10. Based on your inspection of the drum brake components, what are the necessary actions?

_____

_____

11. Before beginning to reassemble the brakes with new parts, compare the old and new parts to make sure the replacement parts are correct.

Instructor's check _____

12. Before beginning to reassemble the brakes, some technicians use masking tape to cover the new linings to prevent contamination from dirt and brake dust. Next, reassemble the drum brake components. Once the brake shoes and hardware are installed, have your instructor check your work.

Instructor's check _____

13. Once installed and checked, you will need to adjust the shoe-to-drum clearance. Describe how to check and adjust the shoe-to-drum clearance. _____

_____

_____

14. Explain why adjusting the shoe-to-drum clearance is an important stop in replacing the rear brake shoes.

_____

_____

_____

Instructor's check _____

15. Once the shoe-to-drum clearance has been adjusted properly, check the operation of the parking brake. Refer to the manufacturer's service information for specific adjustment details. Summarize how to check for proper adjustment. _____

_____

_____

16. Perform a final check of your work, ensuring all brake components are assembled properly, brake shoe adjustment is correct, and the drum is installed properly.

Instructor's check _____

17. Once finished, put away all tools and equipment and clean up the work area.

Instructor's check _____

Lab Worksheet 13-4

Name _____ Station _____ Date _____

Inspect Wheel Cylinders

ASE Education Foundation Correlation

This lab worksheet addresses the following **MLR** task:

5.C.4. Inspect wheel cylinders for leaks and proper operation; remove and replace as needed. **(P-2)**

Description of Vehicle

Year _____ Make _____ Model _____

Engine _____ AT MT CVT PSD (circle which applies)

Procedure

1. Perform a visual inspection of the outside of the brake drum and the backing plate. Look for evidence of fluid leaking around the drum and backing plate. Note your findings. _____
   _____
   _____

2. Remove the brake drum. Describe the type of drum brakes installed on the vehicle.
   _____
   _____

3. Inspect the dust in the drum and on the shoes.

   Does the dust appear wet?   Yes _____   No _____

   If yes, what appears to be the most likely cause of the leak? _____

4. Carefully pull back the edge of one of the wheel cylinder's dust boots.

   Is fluid present in the boot?   Yes _____   No _____

   If yes, what does this indicate? _____

5. Carefully attempt to push a piston back into the cylinder bore.

   Does the piston move inward?   Yes _____   No _____

   If no, what does this indicate? _____

6. Inspect the brake line connection and the bleeder screws at the rear of the wheel cylinders. Describe their appearance. _____
   _____

7. Carefully attempt to loosen the bleeder screws. Do the bleeder screws open easily?

   Yes _____   No _____

   If no, what actions may be required to open the bleeder screws? _____
   _____

8. Based on your inspection what actions are required? _____
   _____
   _____

9. Remove any tools and equipment and clean up the work area.

                                        Instructor's check _____

Lab Worksheet 13-5

Name _____ Station _____ Date _____

Inspect Parking Brake

ASE Education Foundation Correlation

This lab worksheet addresses the following **MLR** task:

5.F.2. Check parking brake system components for wear, binding, and corrosion; clean, lubricate, adjust and/or replace as needed. **(P-2)**

Description of Vehicle

Year _____ Make _____ Model _____

Engine _____ AT MT CVT PSD (circle which applies)

Procedure

1. Describe the type and operation of the parking brake system used on this vehicle. _____
   _____
   _____

2. Raise the vehicle and inspect the parking brake cables and components that are visible. Note your findings.
   _____
   _____

3. Have a helper slowly start to apply the parking brake while you watch the cables. Have the helper stop if the parking brake is very tight. Apply the brake and then release.

   Did the cables move?   Yes _____   No _____

   If no, what may this indicate? _____
   _____

   Did the parking brake release properly?   Yes _____   No _____

4. If the cables do not move or the brake did not release, locate the cable or part that is binding or is frozen. Note your findings. _____

5. If the cables move properly, apply the brake the specified amount according to the manufacturer's service procedures.

   Specified number of clicks _____

   Does the parking brake set properly and hold?   Yes _____   No _____

6. If the parking brake does not fully apply or hold, it may need adjusted. Summarize the method of adjusting the parking brake. _____

   _____

7. Does the parking brake require adjustment?   Yes _____   No _____

8. Based on your inspection, what is the condition of the parking brake system and what service or repairs are necessary? _____

   _____

   _____

9. Remove any tools and equipment and clean up the work area.

                                   Instructor's check _____

CHAPTER 14

Disc Brake System Principles

Review Questions

1. Front disc brakes are _____ on all modern cars and light trucks, and disc brakes are often standard for the _____ brakes as well.

2. Which of the following is not an advantage of disc brakes?
   a. Self-adjusting
   b. Increased fade resistance
   c. Decreased driver effort
   d. Self-cleaning of dust

3. Describe in your own words about basic disc brake operation. _____
   _____
   _____
   _____

4. The _____ is the hydraulic output for the disc brake system.

5. *Technician A* says all disc brake systems use a caliper to push the brake pads out against the brake rotor. *Technician B* says some disc brake systems use multiple calipers to apply the brake pads against the rotor. Who is correct?
   a. Technician A
   b. Technician B
   c. Both A and B
   d. Neither A nor B

6. Which of the following is not a common type of disc brake caliper?
   a. Fixed
   b. Floating
   c. Fixed-sliding
   d. Sliding

355

7. Fixed calipers have at least _____ pistons.

8. In a fixed caliper design, how is the outer brake pad applied against the rotor?
   a. Leverage from the inner caliper piston
   b. Hydraulic pressure
   c. The caliper sliding backward
   d. None of the above

9. The piston seal, sometimes called a _____ seal, seals each piston in the caliper bore.

10. Caliper piston dust boots are _____ type seals, meaning they will expand and contract to cover the piston as the piston moves outward in the bore.

11. Describe how the piston is retracted back into the caliper bore when the brakes are released. _____ _____ _____

12. All of the following statements about floating calipers is correct except:
    a. Floating calipers are smaller and less expensive than fixed calipers.
    b. Floating calipers use Newton's Third Law of Motion to operate.
    c. Floating calipers are often mounted on bushings or sleeves.
    d. Floating calipers use both inboard and outboard pistons.

13. Newton's Third Law of Motion states that for every _____ there is an _____ and _____ reaction.

14. For the _____ caliper to operate correctly, the caliper must be able to move freely on the mounting _____.

15. _____ calipers do not use mounting bolts or pins; instead they are mounted so that they can slide on the steering knuckle.

16. Which of the following is not a common caliper piston construction material?
    a. Aluminum
    b. Plastic
    c. Cast iron
    d. Steel

17. Floating calipers have mounting _____, pins, or bushing bores.

18. Explain the purpose of caliper hardware. _____ _____

19. Label the components of the brake pad shown in Figure 14-1 (A through E).

    a. _____   b. _____

    c. _____   d. _____

    e. _____

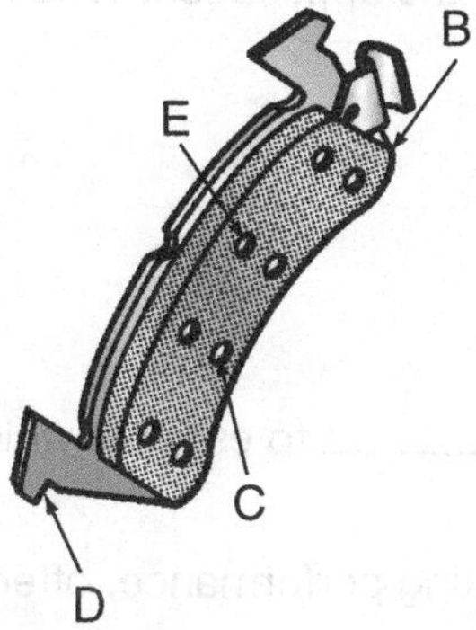

Figure 14-1

20. Many pads also use _____ to help reduce noise.

21. Rotors, also called _____, are mounted to the hub and rotate with the wheel and tire.

22. Label the parts of the brake rotor shown in Figure 14-2 (A through E).

    A. _____   B. _____

    C. _____   D. _____

    E. _____

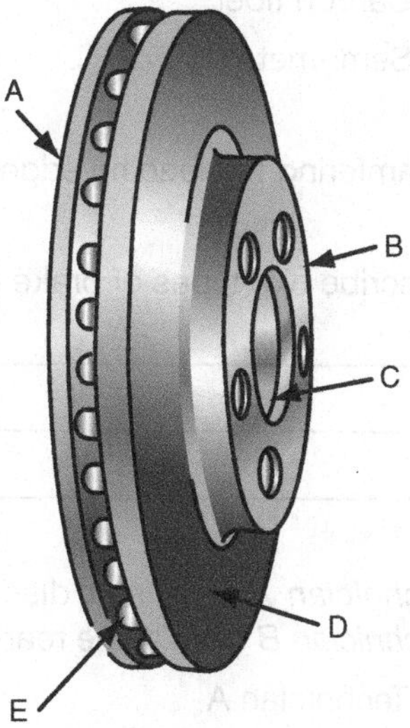

Figure 14-2

23. The center of the rotor is also called the _____.

24. The overall rotor friction surface area is _____, but the contact area of the _____ is small.

25. Explain why nonvented brake rotors are not used on front disc brakes. _____

_____

_____

26. *Technician A* says all disc brake rotors are vented to improve heat dissipation. *Technician B* says rotor vents can be either straight or curved depending on the application. Who is correct?

   a. Technician A

   b. Technician B

   c. Both A and B

   d. Neither A nor B

27. Drilling or slotting rotors allows _____ to escape, which improves braking.

28. Brake _____ is the loss of braking performance, often as a result of heat.

29. Brake pad linings are a _____ between factors such as pad life, noise generation, and cold and hot coefficients of friction.

30. All of the following are brake pad friction material types except:

   a. Organic

   b. Ceramic

   c. Carbon fiber

   d. Semi-metallic

31. Chamfering the leading edges of the brake pads helps decrease brake _____.

32. Describe two types of brake pad wear indicators and their operation. _____

_____

_____

_____

33. *Technician A* says rear disc brake assemblies are similar in operation to front disc brake assemblies. *Technician B* says some rear disc brake systems include the parking brake system. Who is correct?

   a. Technician A

   b. Technician B

   c. Both A and B

   d. Neither A nor B

34. Describe the operation of an integral rear parking brake caliper. _____
_____
_____
_____

35. Rear disc brakes that have a set of small brake shoes enclosed by the brake rotor are called
_____ - _____ - _____ systems.

36. True or False: Some vehicles have electric parking brake motors. _____

37. Describe regenerative braking system operation. _____
_____
_____
_____

Activities

1. Identify the types of the disc brake calipers shown in Figure 14-3, Figure 14-4, and Figure 14-5.

Figure 14-3 _____

Figure 14-4 _____

Figure 14-5 _____

Figure 14-3

Figure 14-4

Figure 14-5

2. Identify the components of the disc brake caliper shown in Figure 14-6 (A through I).

A. _____   B. _____

C. _____   D. _____

E. _____   F. _____

G. _____   H. _____

I. _____

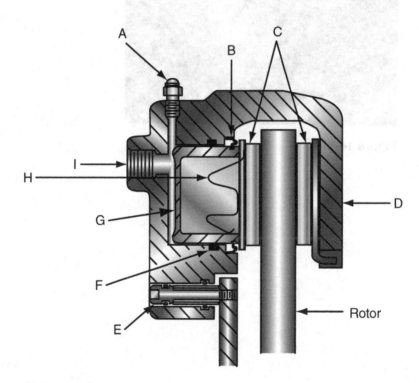

Figure 14-6

Lab Activity 14-1

Name _____ Date _____ Instructor _____

Year _____ Make _____ Model _____

Identify Disc Brake Types

1. Describe the type of front disc brake system used on this vehicle. _____

    _____

2. Caliper type: Fixed _____   Floating _____   Sliding _____

3. Number of caliper pistons   _____

    Caliper piston construction material _____

4. Rotor type: Floating _____   Trapped _____

5. Describe the type of rear disc brake system used on this vehicle. _____

    _____

6. Caliper type: Fixed _____   Floating _____   Sliding _____

7. Number of caliper pistons   _____

    Caliper piston construction material _____

8. Rotor type: Floating _____   Trapped _____

9. Describe the type of parking brake system used. _____

    _____

CHAPTER 15

Disc Brake System Inspection and Service

Review Questions

1. Performing a brake job means more than just installing a set of brake _____.

2. True or False: Brake work is not a common service performed by technicians. _____

3. Improper use of tools and taking shortcuts can lead not only to _____ damage, but can also cause personal _____.

4. It is important to always use the proper _____ for the job.

5. What is the purpose of the tool shown in Figure 15-1?

Figure 15-1

6. To measure brake rotor wear, a rotor _____ is used.

7. A _____ is used to measure brake rotor runout.

8. List three precautions for working on the disc brake system. _____
   _____
   _____
   _____

9. Wearing _____ helps protect your hands and can help prevent getting new components dirty during installation.

10. Summarize three safety precautions for working on the brake system. _____
    _____
    _____
    _____
    _____

11. List five methods of ensuring that brake work is performed safely and properly.
    a. _____
    b. _____
    c. _____
    d. _____
    e. _____

12. *Technician A* says replacing the brake pads can cause the brake pedal to feel different and can affect stopping performance. *Technician B* says flushing and filling the hydraulic system with a different brake fluid can affect brake pedal feel. Who is correct?
    a. Technician A
    b. Technician B
    c. Both A and B
    d. Neither A nor B

13. Because of the way in which disc brakes operate, they are prone to _____ issues.

14. Describe how noise is generated by the disc brake system. _____
    _____
    _____
    _____

15. What is the purpose of the component shown in Figure 15-2?

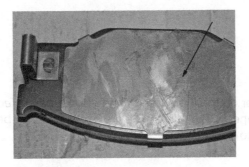

Figure 15-2

16. A rhythmic sound like a wire brush against metal can be caused by rotor _____ or
_____ variation.

17. List five possible causes of disc brake noise.

a. _____

b. _____

c. _____

d. _____

e. _____

18. All of the following can cause disc brake pulsation concerns except:

a. Out-of-round rotor

b. Excessive thickness variation

c. Hub runout

d. Loose wheel bearing

19. The most common causes of disc brake _____ are restricted hydraulic hoses and
_____ caliper pistons.

20. Worn and loose control arm _____ can cause the vehicle to pull when braking.

21. Describe how to check for restricted hoses and seized calipers. _____

_____

_____

22. Brake _____ refers to a brake that applies with too much braking force, causing the wheel
to lock up easily.

23. Brake _____ occurs when a brake remains applied after the brake pedal is released.

24. Which of the following is not a common disc brake complaint?

    a. Noise

    b. Pulsation

    c. Pulling

    d. Firm pedal feel

25. A vehicle has uneven pad wear on the left and right side brake pads. *Technician A* says a caliper may be sticking and not releasing properly. *Technician B* says excessive rust between the pads and the caliper bracket may be the cause. Who is correct?

    a. Technician A

    b. Technician B

    c. Both A and B

    d. Neither A nor B

26. Begin your brake inspection with a _____ with the driver of the vehicle.

27. Both the red _____ light and the _____ light should illuminate during engine start and bulb check and then go out after a few seconds.

28. A leaking caliper must be either rebuilt or _____.

29. Inspect the hoses for signs of _____ or contact with other components.

30. Brake pads should be replaced when the lining reaches approximately _____ (inch or mm) in thickness.

31. *Technician A* says a pad lining will wear faster once it is half worn out. *Technician B* says a pad will wear at the same rate regardless of how much wear is on the linings. Who is correct?

    a. Technician A

    b. Technician B

    c. Both A and B

    d. Neither A nor B

32. Uneven pad wear can indicate that the pads are not properly and are not fully releasing from the applied position against the rotor.

33. Before starting to _____ the brakes, first ensure you understand how the brake is assembled.

34. What action is being performed in Figure 15-3?

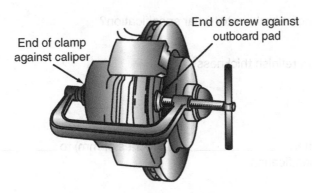

End of clamp against caliper

End of screw against outboard pad

Figure 15-3

35. Do not allow the _____ to hang on the brake, as this can damage the hose.

36. When you are ready to reassemble the brakes, use a small amount of brake _____ on the caliper pins.

37. Before you start and move the vehicle after the pads have been replaced, _____ the brake pedal several times until the pedal pumps up and is _____.

38. Never let the _____ be the first person to drive the vehicle after the brakes have been serviced.

39. Describe pad break-in and why it is important for the pads and rotors. _____
_____
_____
_____

40. Brake rotors should be free of:
    a. Deep scoring
    b. Glazing
    c. Cracks
    d. Deep scoring, glazing, and cracks

41. *Technician A* says that a rotor may rust from the inside out, causing the rotor to separate through the vents. *Technician B* says that the inner surface of the rotor should be closely inspected for rust. Who is correct?
    a. Technician A
    b. Technician B
    c. Both A and B
    d. Neither A nor B

42. Which of the following is not a brake rotor wear specification?

    a. Maximum thickness

    b. Machine-to or minimum refinish thickness

    c. Runout

    d. Parallelism

43. The machine-to spec is often _____ (inches or mm) to _____ (inches or mm) larger than the discard specification.

44. What measurement is shown in Figure 15-4 and why is it important?

    _____

    _____

    _____

Figure 15-4

45. Why do some vehicles use on-car brake rotor machining? _____

    _____

    _____

    _____

46. Explain how the caliper piston is retracted on an integral parking brake caliper.

    _____

    _____

47. Explain the final checks you should perform once the brake pads have been replaced.

    _____

    _____

    _____

Activities

1. Place the following brake pad service steps in proper order.

   a. Measure rotor thickness

   b. Loosen and remove caliper pins

   c. Set caliper piston

   d. Remove pads

   e. Remove caliper and hang from wire

   f. Reinstall and torque

   g. Lube pad guides

   h. Lube caliper pins

2. Match the following complaints with their most likely causes.

   Pulsation

   Squealing

   Drag

   Abnormal pad wear

   Pull

   Restricted brake hose

   Thickness variation

   Stuck caliper piston

   Binding caliper pins

   Hydraulic leak

Lab Worksheet 15-1

Name _____ Station _____ Date _____

Inspect Caliper

ASE Education Foundation Correlation

This lab worksheet addresses the following **MLR** task:

5.D.1. Remove and clean caliper assembly; inspect for leaks and damage/wear; determine necessary action. **(P-1)**

Description of Vehicle

Year _____ Make _____ Model _____

Engine _____ Disc brake type _____

Procedure

Safely raise and support the vehicle.

1. Refer to the manufacturer's service information for inspection and service procedures for the disc brake system. Summarize these procedures. _____

_____

_____

2. Perform a visual inspection of the caliper assembly. Looks for signs of fluid leaks around the caliper piston dust boot, brake hose, and bleeder screw. Note your findings. _____

_____

_____

3. What types of concerns can be caused by a brake fluid leak from the caliper? _____

_____

_____

4. Remove the caliper mounting bolts. Remove the caliper from the steering knuckle. Clean the caliper and inspect the dust boot for damage. Inspect the caliper mountings for damage. Note your findings.

_____

_____

5. What concerns can be caused by a damaged piston dust boot? _____
_____
_____

6. Based on your inspection, what is the condition of the caliper? _____
_____
_____

7. Once finished, put away all tools and equipment and clean up the work area.

Instructor's check _____

Lab Worksheet 15-2

Name _____ Station _____ Date _____

Inspect Caliper Mount and Hardware

ASE Education Foundation Correlation

This lab worksheet addresses the following **MLR** task:

5.D.2. Inspect caliper mounting and slides/pins for proper operation, wear, and damage; determine necessary action. **(P-1)**

Description of Vehicle

Year _____ Make _____ Model _____

Engine _____ Disc brake type _____

Procedure

Safely raise and support the vehicle.

1. Refer to the manufacturer's service information for inspection and service procedures for the disc brake system. Summarize these procedures. _____

    _____

    _____

2. With the caliper removed, inspect the caliper mounting pins, bolts, and bracket. Note your findings.

    _____

    _____

3. Remove the pads and inspect the pad hardware and guides. Note any problems. _____

    _____

    _____

4. What concerns can result from worn, seized, or damaged caliper mounting hardware? _____

    _____

    _____

5. Based on your inspection, what is the condition of the caliper hardware? _____

_____

_____

6. Once finished, put away all tools and equipment and clean up the work area.

Instructor's check _____

Lab Worksheet 15-3

Name _____ Station _____ Date _____

Inspect Pads

ASE Education Foundation Correlation

This lab worksheet addresses the following **MLR** task:

5.D.3. Remove, inspect, and/or replace brake pads and retaining hardware; determine necessary action. **(P-1)**

Description of Vehicle

Year _____ Make _____ Model _____

Engine _____ Disc brake type _____

Procedure

Safely raise and support the vehicle.

1. Refer to the manufacturer's service information for specific procedures to remove and service the disc brakes for this vehicle.

    Instructor's check _____

2. Remove the disc brake caliper mounting bolts. Carefully remove the caliper from its mount. Use a bungee cord or wire hanger to support the caliper once removed—*do not let the caliper hang by the brake hose*.

    Instructor's check _____

3. Remove the brake pads and any hardware. Inspect the pads and related hardware for wear and damage. Note your findings. _____

    _____

    _____

    _____

4. Using a brake lining thickness gauge or similar tool, measure the brake pad linings at their thinnest points. Compare the measurement with the brake pad wear limits in the service information.

    Pad wear limit specification _____

    Measured minimum thickness _____

5. Inspect the caliper for signs of leaks and damage. Inspect the caliper bracket for signs of wear or damage. Note your findings. _____

_____

_____

6. Inspect the brake hardware, this includes guides, shims, clips, and other parts that hold the pads in place or prevent noise. Note your findings. _____

_____

_____

7. Based on your inspection, what are the necessary actions? _____

_____

_____

8. Once finished, put away all tools and equipment and clean up the work area.

Instructor's check _____

Lab Worksheet 15-4

Name _____ Station _____ Date _____

Inspect Pads Part 2

ASE Education Foundation Correlation

This lab worksheet addresses the following **MLR** task:

5.D.3. Remove, inspect, and/or replace brake pads and retaining hardware; determine necessary action. **(P-1)**

Description of Vehicle

Year _____ Make _____ Model _____

Engine _____ AT MT CVT PSD (circle which applies)

Identify Pad Wear

1. Describe the type of disc brake system being inspected. _____
_____

2. Check for signs of abnormal pad and rotor wear by performing a visual inspection. Note your findings.
_____
_____

3. Remove the brake caliper from the knuckle or bracket to gain access to the pads. Describe what steps you used to remove the caliper. _____
_____

4. Carefully remove the pads. Inspect the lining areas and note any of the following conditions:

   a. Excessive wear   Yes _____   No _____

   Note minimum lining thickness found _____

   Note minimum lining specification _____

    b. Cracks   Yes _____   No _____

    c. Glazing   Yes _____   No _____

    d. Uneven wear   Yes _____   No _____

    If yes, describe the wear _____

_____

    e. Other _____

5. Based on your inspection, what actions are necessary? _____

_____

_____

6. Remove any tools and equipment and clean up the work area.

                                 Instructor's check _____

Lab Worksheet 15-5

Name _____ Station _____ Date _____

Service Pads

ASE Education Foundation Correlation

This lab worksheet addresses the following **MLR** task:

5.D.4. Lubricate and reinstall caliper, brake pads, and related hardware; seat brake pads and inspect for leaks. **(P-1)**

Description of Vehicle

Year _____ Make _____ Model _____

Engine _____ Disc brake type _____

Procedure

Safely raise and support the vehicle.

1. Once you have removed and inspected the caliper, pads, and hardware, determine what parts need to be replaced. Record your recommendations. _____

   _____

   _____

2. When replacing brake pads and their hardware, lubrication may be required on some of the surfaces. Refer to the manufacturer's service information and determine what parts should have lubricant applied to them. Note your findings. _____

   _____

3. Refer to the manufacturer's service information for what type of lubricant should be used on the brake part. Note your findings. _____

4. What can result if the wrong type of lubricant is used on brake components. _____

   _____

5. Apply lubricant at required.

   Instructor's check _____

6. Reinstall the brake pads and hardware. Ensure all the parts fit together properly and that the pads are able to slide within the guides.

Instructor's check _____

7. Reinstall the caliper. Align the mounting bolts and torque to specifications.

Torque specifications _____

Instructor's check _____

8. Apply the brake pedal several times until the pedal becomes firm. Check for brake fluid leaks from the caliper.

Instructor's check _____

9. Some brake pads require setting into place once installed. Refer to the service information for instructions on seating the pads. Note your findings. _____

_____

10. Once new brake pads are installed, they should be burnished or bedded to provide optimal pad and rotor life and reduced operating noise. Refer to the service information or instructions that came with the new pads on break-in procedures. Record the procedure. _____

_____

11. What concerns can result from not properly breaking-in a new set of brake pads?

_____

_____

12. Once finished, put away all tools and equipment and clean up the work area.

Instructor's check _____

Lab Worksheet 15-6

Name _____ Station _____ Date _____

Inspect Rotor

ASE Education Foundation Correlation

This lab worksheet addresses the following **MLR** task:

5.D.5. Clean and inspect rotor and mounting surface, measure rotor thickness, thickness variation, and lateral runout; determine necessary action. **(P-1)**

Description of Vehicle

Year _____ Make _____ Model _____

Engine _____ AT MT CVT PSD (circle which applies)

Procedure

1. Locate and record the rotor specs.

   Nominal thickness _____   Minimum thickness _____

   Machine-to _____   Parallelism _____

2. Place the rotor micrometer in the middle of the rotor's friction surface and note the reading.

   Micrometer reading _____

3. How does the rotor measurement compare to the minimum thickness spec?

   Over specification _____   Under specification _____   Within specification _____

4. Next, measure the rotor thickness at eight places around the friction surface and record your readings.

   _____   _____   _____   _____

   _____   _____   _____   _____

   Subtract the smallest number from the largest to find the parallelism difference.

   Parallelism measured _____

5. How does this measurement compare to the spec?

Over specification _____   Under specification _____   Within specification _____

6. What causes excessive rotor parallelism? _____
_____

7. Based on your inspection, what is the condition of this rotor, and should it be placed back into service?
_____
_____

8. Remove any tools and equipment and clean up the work area.

Instructor's check _____

Lab Worksheet 15-7

Name _____ Station _____ Date _____

Measure Rotor Runout

ASE Education Foundation Correlation

This lab worksheet addresses the following **MLR** task:

5.D.5. Clean and inspect rotor and mounting surface, measure rotor thickness, thickness variation, and lateral runout; determine necessary action. **(P-1)**

Description of Vehicle

Year _____ Make _____ Model _____

Engine _____ AT MT CVT PSD (circle which applies)

Procedure

1. Locate and record the rotor specs.   Total runout _____

2. Install a dial indicator to measure rotor runout. Turn the rotor slowly and note the total needle movement.

   Total runout measured _____

3. How does this measurement compare to specs?

   Over specification _____ Under specification _____ Within specification _____

4. What are three possible causes for excessive rotor runout? _____

   _____

   _____

5. If the total indicated runout is greater than specification, mark the location of the rotor to the hub, remove the rotor and clean both the rotor and hub mounting surfaces. Reinstall the rotor 180 degrees from its original position, and re-measure runout.

   Runout measured _____

6. If the runout decreased, what does this indicate? _____

_____

7. If the runout remained excessive, remove the rotor and measure the runout of the hub.

   Runout measured _____

8. What is indicated by this measurement? _____

_____

9. Based on your inspection, what is the reason for the runout and what should be done to correct the problem?

_____

_____

_____

_____

10. Remove any tools and equipment and clean up the work area.

Instructor's check _____

Lab Worksheet 15-8

Name _____ Station _____ Date _____

Remove and Reinstall Rotor

ASE Education Foundation Correlation

This lab worksheet addresses the following **MLR** task:

5.D.6. Remove and reinstall/replace rotor. **(P-1)**

Description of Vehicle

Year _____ Make _____ Model _____

Engine _____ Disc brake type _____

Procedure

Safely raise and support the vehicle.

1. Refer to the manufacturer's service information for specific procedures to remove and service the disc brakes for this vehicle. Summarize the procedures. _____

_____

Instructor's check _____

2. Remove the disc brake caliper mounting bolts. Carefully remove the caliper from its mount. Use a bungee cord or wire hanger to support the caliper once removed—*do not let the caliper hang by the brake hose*.

Instructor's check _____

3. Brake rotors may be floating, able to slide off the hub, or trapped. Trapped rotors require removing the hub or bearings to remove the rotor. Describe the type of rotor for this vehicle. _____

_____

4. Remove the brake rotor following the procedures in the service information.

Instructor's check _____

5. Inspect the rotor mounting surface on the hub. Check for excessive rust or corrosion on the hub that can prevent the rotor from seating properly. Note your findings. _____

_____

_____

6. What types of concerns can be caused by the rotor not seating properly on the hub?

_____

_____

_____

7. Clean the hub as necessary before installing the rotor.

Instructor's check _____

8. Compare the new rotor with the original. Make sure the mounting face and hub centerbore hole match.

Instructor's check _____

9. New rotors often are covered in a protective film to prevent rusting during shipping. This protectant needs to be removed before installing the rotor. Use brake cleaner or a similar product to remove the protectant from the rotor.

Instructor's check _____

10. Install the new rotor onto the hub. Ensure it fits correctly. Thread a lug nut onto a stud to help keep the rotor in place and make reassembling the caliper and pads easier. Reinstall the pads, hardware, and brake caliper. Torque the caliper mounting bolts to specifications.

Torque specifications _____

Instructor's check _____

11. Apply the brake pedal several times until the pedal becomes firm. Check for brake fluid leaks from the caliper.

Instructor's check _____

12. Some brake pads require setting into place once installed. Refer to the service information for instructions on seating the pads. Note your findings. _____

_____

13. Once new brake pads are installed, they should be burnished or bedded to provide optimal pad and rotor life and reduced operating noise. Refer to the service information or instructions that came with the new pads on break-in procedures. Record the procedure. _____

14. What concerns can result from not properly breaking in a new set of brake pads?

    _____

    _____

15. Once finished, put away all tools and equipment and clean up the work area.

Instructor's check _____

13. Once new brake pads are installed, they should be burnished or bedded to provide optimal pad and rotor life and reduced operating noise. Refer to the service information or instructions that came with the new pads on break-in procedures. Record the procedure. _____

14. What concerns can result from not properly breaking in a new set of brake pads?

_____

_____

15. Once finished, put away all tools and equipment and clean up the work area.

_____ Instructor's check.

Lab Worksheet 15-9

Name _____ Station _____ Date _____

On-Car Lathe

ASE Education Foundation Correlation

This lab worksheet addresses the following **MLR** task:

5.D.7. Refinish rotor on vehicle; measure final rotor thickness and compare with specifications. **(P-1)**

Description of Vehicle

Year _____ Make _____ Model _____

Engine _____ Disc brake type _____

Procedure

Safely raise and support the vehicle.

1. Refer to the manufacturer's service information for specific procedures to remove and service the disc brakes for this vehicle. Summarize the procedures. _____

_____

_____

Instructor's check _____

2. Remove the disc brake caliper mounting bolts. Carefully remove the caliper from its mount. Use a bungee cord or wire hanger to support the caliper once removed—*do not let the caliper hang by the brake hose.*

Instructor's check _____

3. Locate and record the brake rotor machine to, discard, and runout specifications.

Machine to specification _____

Discard specification _____

Rotor runout specification _____

4. Use a brake rotor micrometer to measure the brake rotor thickness. Record your results.

   Measured thickness _____

5. Set up a dial indicator and measure rotor runout.

   Measured runout _____

6. Install the on-car brake lathe according to the lathes operating instructions.

   Instructor's check _____

7. Compensate for runout following the operating instructions.

   Instructor's check _____

8. Once set, machine the rotor. When complete, determine if a second cut is needed.

   Instructor's check _____

9. If needed, perform a final cut of the rotor. When complete, measure runout and thickness

   Measured runout _____

   Measured thickness _____

10. Remove the lathe and clean the brake rotor. Reinstall the brake caliper and pads following the manufacturer's service information.

    Instructor's check _____

11. Once finished, put away all tools and equipment and clean up the work area.

    Instructor's check _____

Lab Worksheet 15-10

Name _____ Station _____ Date _____

Off-Car Lathe

ASE Education Foundation Correlation

This lab worksheet addresses the following **MLR** task:

5.D.8. Refinish rotor off vehicle; measure final rotor thickness and compare with specifications. **(P-1)**

Description of Vehicle

Year _____ Make _____ Model _____

Engine _____ Disc brake type _____

Procedure

Safely raise and support the vehicle.

1. Refer to the manufacturer's service information for specific procedures to remove and service the disc brakes for this vehicle. Summarize the procedures. _____

   _____

   _____

   Instructor's check _____

2. Remove the disc brake caliper mounting bolts. Carefully remove the caliper from its mount. Use a bungee cord or wire hanger to support the caliper once removed—*do not let the caliper hang by the brake hose*.

   Instructor's check _____

3. Locate and record the brake rotor machine to, discard, and runout specifications.

   Machine to specification _____

   Discard specification _____

   Rotor runout specification _____

4. Use a brake rotor micrometer to measure the brake rotor thickness. Record your results.

   Measured thickness _____

5. Set up a dial indicator and measure rotor runout.

   Measured runout _____

6. Remove the brake rotor from the vehicle and clean the rotor mounting surfaces. Explain the steps necessary to remove this rotor. _____

   _____

   _____

   Instructor's check _____

7. Refer to the lathe manufacturers set up instructions and mount the rotor on the brake lathe.

   Instructor's check _____

8. Perform a fast cut of the rotor. When complete, determine if a second cut is needed.

   Instructor's check _____

9. Perform a final cut of the rotor. When complete, finish the rotor surface in a nondirectional pattern. Clean the rotor and thickness

   Measured thickness _____

10. Reinstall the brake caliper and pads following the manufacturer's service information.

    Instructor's check _____

11. Once finished, put away all tools and equipment and clean up the work area.

    Instructor's check _____

Lab Worksheet 15-11

Name _____ Station _____ Date _____

Integral Parking Brake Caliper Service

ASE Education Foundation Correlation

This lab worksheet addresses the following **MLR** task:

5.D.9. Retract and re-adjust caliper piston on an integral parking brake system. **(P-2)**

Description of Vehicle

Year _____ Make _____ Model _____

Engine _____ Disc brake type _____

Procedure

Safely raise and support the vehicle.

1. Refer to the manufacturer's service information for specific procedures to remove and service the disc brakes for this vehicle. *Important Note: If the vehicle has an electrically operated parking brake, DO NOT attempt to seat the piston without following the vehicle manufacturer's service procedures. Serious damage to the caliper can result from improper service procedures.* Summarize the procedures. _____

   _____

   _____

   _____

   Instructor's check _____

2. Remove the disc brake caliper mounting bolts. Carefully remove the caliper from its mount. Use a bungee cord or wire hanger to support the caliper once removed—*do not let the caliper hang by the brake hose.*

   Instructor's check _____

3. Obtain the piston retraction tool and read the instructions on how to setup and use the tool. Explain why a special tool is necessary for retracting the piston for this caliper. _____

   _____

   _____

4. Install the piston retraction tool and slowly begin to turn the piston back into its bore. The piston should turn with little resistance and fully seat into the caliper. Note any problems encountered retracting the piston.

_____

_____

5. Check for signs of brake fluid leaks around the caliper piston seal. If any fluid is present, what is the necessary action? _____

_____

6. Once the piston is fully retracted, reinstall the pads and caliper. Note any gap between the piston and the inboard brake pad. If a gap is present, the piston will need to be adjusted. What concerns can result from improper adjustment of the piston and inboard pad on this brake system? _____

_____

_____

7. Refer to the manufacturer's service information about how to adjust the rear caliper. Summarize the procedure. _____

_____

8. Perform the rear caliper adjustment and recheck the fit of the rear caliper and pads.

Instructor's check _____

9. Once the rear caliper adjustment is correct, recheck your work and torque all fasteners to specifications.

Instructor's check _____

10. Check brake operation by pressing the brake pedal several times, confirming the pedal feel is good. Operate the parking brake several times and make sure the parking brake holds the vehicle in place.

Instructor's check _____

11. Once finished, put away all tools and equipment and clean up the work area.

Instructor's check _____

Lab Worksheet 15-12

Name _____ Station _____ Date _____

Inspect Pad Wear Indicator

ASE Education Foundation Correlation

This lab worksheet addresses the following **MLR** task:

5.D.10. Check brake pad wear indicator; determine necessary action. **(P-1)**

Description of Vehicle

Year _____ Make _____ Model _____

Engine _____ Disc brake type _____

Procedure

Safely raise and support the vehicle.

1. Refer to the manufacturer's service information to inspect the disc brake wear indicator for this vehicle. Summarize the procedures. _____

_____

_____

Instructor's check _____

2. Locate the brake pad minimum thickness specification.

Minimum thickness _____

Instructor's check _____

3. Inspect the wear indicator. This may be a simple metal tab or an electronic sensor built into the pad lining.

Wear indicator type _____

4. Use a brake lining thickness gauge or tire tread depth gauge to measure brake lining thickness. Record your findings.

RF inner _____   outer _____   LF inner _____   outer _____

RF inner _____   outer _____   LF inner _____   outer _____

5. Based on your inspection, what is the necessary action? _____

_____

6. Once finished, put away all tools and equipment and clean up the work area.

Instructor's check _____

Lab Worksheet 15-13

Name _____ Station _____ Date _____

Pad Break-In

ASE Education Foundation Correlation

This lab worksheet addresses the following **MLR** task:

5.D.11. Describe importance of operating vehicle to burnish/break-in replacement brake pads according to manufacturer's recommendation.**(P-1)**

Description of Vehicle

Year _____ Make _____ Model _____

Engine _____ Front/Rear brake type _____

Procedure

1. Describe how friction is used by the brake system. _____
_____
_____

2. Explain the factors that affect the friction between the brake pad linings and the rotor surfaces.
_____
_____
_____

3. Describe what concerns can result from too much or too little friction between the brake pad linings and the rotors._____
_____
_____

4. Describe how the manufacturer recommends breaking in or bedding new brake pads.

_____

_____

_____

5. Describe what concerns can result from not properly breaking in or bedding new brake linings.

_____

_____

_____

CHAPTER 16

Antilock Brakes, Stability Control, and Power Assist System

Review Questions

1. Antilock brake systems use a _____ and _____ to monitor wheel speeds.

2. Vehicle stability control systems are designed to work with the antilock brake system to reduce _____ and _____.

3. *Technician A* says ABS is designed to decrease stopping distances in all braking situations. *Technician B* says ABS is designed to allow the driver to maintain control by preventing wheel lock. Who is correct?

   a. Technician A

   b. Technician B

   c. Both A and B

   d. Neither A nor B

4. ABS prevents wheel lock by which of the following actions?

   a. Applying reduced pressure to the wheel brake cylinders

   b. Applying increased pressure to the wheel brake cylinders

   c. Applying, holding, and releasing the pressure to the wheel brake cylinders

   d. None of the above

5. ABS and traction control systems cannot overcome which of the following factors?

   a. Insufficient traction between the tire and the ground

   b. Inoperative or faulty brake system components

   c. Gross driver negligence

   d. All of the above

6. The amount of traction between the tire and the ground is called tire _____.

7. _____ monitor the rotational speed of each wheel.

8. Which of the following components receives input and controls ABS operation?

    a. Wheel speed sensors

    b. EBCM

    c. Electro-hydraulic control unit

    d. Powertrain control module

9. EBCM stands for: _____

    _____

10. List the three modes in which the EBCM may command the hydraulic control unit.

    _____

    _____

11. The _____ usually contains electric motors, solenoids, and valves that control the flow of brake fluid.

12. *Technician A* says wheel speed sensors can produce either analog or digital signals. *Technician B* says all wheel speed sensors produce analog AC voltage signals. Who is correct?

    a. Technician A

    b. Technician B

    c. Both A and B

    d. Neither A nor B

13. Vehicles with ABS but without traction or stability control are being discussed: *Technician A* says the ABS only monitors wheel speed when the brakes are applied. *Technician B* says the ABS monitors wheel speed whenever the vehicle is moving. Who is correct?

    a. Technician A

    b. Technician B

    c. Both A and B

    d. Neither A nor B

14. Isolate mode is also called:

    a. Release mode

    b. Hold mode

    c. Reapply mode

    d. Pressure increase mode

15. During release mode, the EBCM commands pressure release so the wheel can _____.

16. Explain in detail the sequence of steps the ABS uses to control wheel lockup.

    _____

    _____

    _____

    _____

17. Never open a _____ line or _____ screw with the ignition on, as this can cause injury from the release of high-pressure fluid.

18. Integral ABS units use high-pressure _____, which must be discharged before any service is performed.

19. The first step in diagnosing an ABS concern is to _____ and _____ the customer complaint.

20. Explain why an inspection of the brake system components is important when diagnosing an ABS problem.

    _____

    _____

    _____

21. To flush and bleed the ABS hydraulic system, a _____ may be required.

22. Describe the purpose of the traction control system. _____

    _____

    _____

23. List five requirements of electronic stability control systems.

    a. _____

    b. _____

    c. _____

    d. _____

    e. _____

24. Describe how to relearn or reset a steering angle sensor.

    _____

    _____

    _____

    _____

25. There are two types of external power brake assist systems, _____ assist and _____ assist.

26. Vacuum is air pressure that is less than _____ pressure.

27. Explain the purpose of the vacuum check valve. _____

    _____

28. Some engines use a _____ pump to supply vacuum to the booster.

29. Explain how pressure differential is used to provide power assist in a vacuum booster.

_____

_____

_____

_____

30. Describe how atmospheric pressure, vacuum, and booster diaphragm size contribute to the amount of power assist generated by the vacuum booster. _____

_____

_____

_____

_____

31. Hydraulic power brake boosters may use fluid from the pump _____ or an _____ pump.

32. _Technician A_ says that a loose or missing power steering belt can affect the operation of a Hydro-boost system. _Technician B_ says a faulty power steering pump can cause excessive brake pedal effort on a Hydro-boost system. Who is correct?

    a. Technician A

    b. Technician B

    c. Both A and B

    d. Neither A nor B

33. A vehicle with a vacuum power assist system has a very hard brake pedal and requires excessive effort to stop. All of the following can be the problem except:

    a. A leaking power booster diaphragm

    b. A misadjusted booster pushrod

    c. A blocked vacuum supply hose

    d. A leaking vacuum supply hose

34. When testing a vacuum power assist unit, _Technician A_ says the brake pedal should slowly sink to the floor under light pressure when the engine is started. _Technician B_ says the brake pedal should drop a couple of inches under light pressure once the engine is started. Who is correct?

    a. Technician A

    b. Technician B

    c. Both A and B

    d. Neither A nor B

35. *Technician A* says servicing brake systems on some hybrid vehicles requires following special service precautions and procedures. *Technician B* says some hybrid vehicles may apply the brakes even if the engine and ignition system are turned off. Who is correct?

    a. Technician A

    b. Technician B

    c. Both A and B

    d. Neither A nor B

Activities

1. Describe the purpose of the component shown in Figure 16-1.

_____

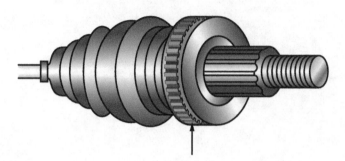

Figure 16-1

2. Label the components of the ABS system shown in Figure 16-2 (A through G).

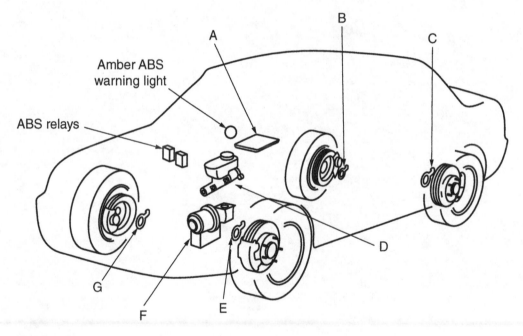

Figure 16-2

3. Label the components of the integral ABS system shown in Figure 16-3 (A through G).

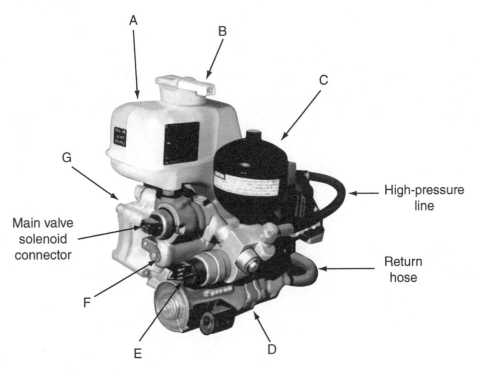

A

B

C

G

High-pressure
line

Main valve
solenoid
connector

Return
hose

F

E

D

Figure 16-3

4. Label the parts of the vacuum booster shown in Figure 16-4 (A through G).

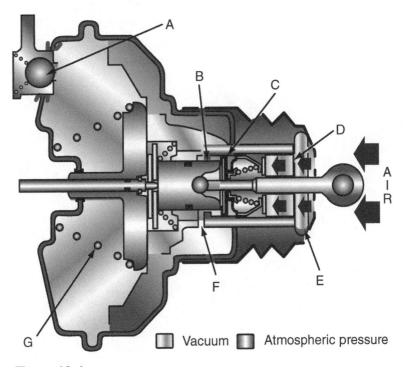

A

B

C

D

A
I
R

F

E

G

Vacuum   Atmospheric pressure

Figure 16-4

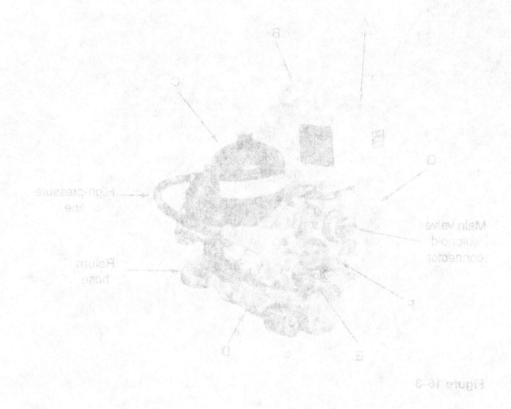

High-pressure line

Return hose

Main valve solenoid connector

Figure 16-3

4. Label the parts of the vacuum booster shown in Figure 16-4 (A through G).

☐ Vacuum   ☐ Atmospheric pressure

Figure 16-4

Lab Worksheet 16-1

Name _____ Station _____ Date _____

Check Booster Operation

ASE Education Foundation Correlation

This lab worksheet addresses the following **MLR** task:

5.E.1. Check brake pedal travel with, and without, engine running to verify proper power booster operation. **(P-2)**

Description of Vehicle

Year _____ Make _____ Model _____

Engine _____ AT MT CVT PSD (circle which applies)

Procedure

1. Perform a visual inspection of the vacuum booster and vacuum hose. Note your findings.

   _____

2. With the engine off, pump the brake pedal until the vacuum reserve is depleted. With your foot holding the brake pedal down, start the engine and note how the brake pedal responds.

   _____

3. Next, pump the brake pedal several times and note how it feels.

   Very hard to depress   Yes _____   No _____

   Moves down easily   Yes _____   No _____

4. With the brake depressed, listen for a hissing noise inside the vehicle.

   Is there a hiss present with the brake applied?   Yes _____   No _____

   What may a hiss with the brake applied indicate? _____

5. Turn the engine off and press the brake pedal down once. Does the pedal move down with normal pressure applied?   Yes _____   No _____

   If the pedal is very hard to press, what might this indicate? _____

6. Based on your inspection, what is the condition of the brake booster? _____

_____

_____

7. Remove any tools and equipment and clean up the work area.

Instructor's check _____

Lab Worksheet 16-2

Name _____ Station _____ Date _____

ID Power Assist Components

ASE Education Foundation Correlation

This lab worksheet addresses the following **MLR** task:

5.E.2. Identify components of the brake power assist system (vacuum and hydraulic); check vacuum supply (manifold or auxiliary pump) to vacuum-type power booster. **(P-1)**

Description of Vehicle

There are three ways to provide power brake assist; vacuum assist, hydraulic assist, and integral ABS assist.

Vehicle 1: Year _____ Make _____ Model _____

Identify the type of assist used on this vehicle. _____

Vehicle 2: Year _____ Make _____ Model _____

Identify the type of assist used on this vehicle. _____

Vehicle 3: Year _____ Make _____ Model _____

Identify the type of assist used on this vehicle. _____

Procedure

1. On a vehicle with vacuum assist, identify the following components:

   a. Vacuum assist unit _____

   b. Vacuum supply hose _____

   c. Vacuum check valve _____

2. On a vehicle with hydraulic assist, identify the following components:

   a. Power assist unit   _____

   b. Power steering pump   _____

   c. High-pressure supply hose _____

   d. Return hose   _____

3. On a vehicle with an integral ABS unit, identify the following components:

   a. High-pressure accumulator _____

   b. Hydraulic unit   _____

Instructor's check _____

Lab Worksheet 16-3

Name _____ Station _____ Date _____

Check Vacuum

ASE Education Foundation Correlation

This lab worksheet addresses the following **MLR** task:

5.E.2. Identify components of the brake power assist system (vacuum and hydraulic); check vacuum supply (manifold or auxiliary pump) to vacuum-type power booster. **(P-1)**

Description of Vehicle

Year _____ Make _____Model _____

Engine _____ AT MT CVT PSD (circle which applies)

Procedure

1. Perform a visual inspection of the vacuum booster and vacuum hose. Note your findings.

   _____

2. With the engine off, pump the brake pedal until the vacuum reserve is depleted. With your foot holding the brake pedal down, start the engine and note how the brake pedal responds.

   _____

3. Next, pump the brake pedal several times and note how it feels.

   Very hard to depress   Yes _____   No _____

   Moves down easily   Yes _____   No _____

4. With the brake depressed, listen for a hissing noise inside the vehicle.

   Is there a hiss present with the brake applied?   Yes _____   No _____

   What may a hiss with the brake applied indicate? _____

5. Turn the engine off and press the brake pedal down once. Does the pedal move down with normal pressure applied?   Yes _____   No _____

   If the pedal is very hard to press, what might this indicate? _____

6. Disconnect the vacuum hose to the brake booster and install a vacuum gauge onto the hose. Start the engine and note the vacuum reading.

   Vacuum reading at idle _____

7. Refer to the service information for specifications about the vacuum supply to the power brake booster.

   Vacuum specification _____

8. Based on your inspection, is the vacuum supply to the boost acceptable?

   Yes _____   No _____

9. Remove any tools and equipment and clean up the work area.

   Instructor's check _____

Lab Worksheet 16-4

Name _____ Station _____ Date _____

ID Components

ASE Education Foundation Correlation

This lab worksheet addresses the following **MLR** task:

5.G.1. Identify traction control/vehicle stability control system components. **(P-3)**

Description of Vehicle

Year _____ Make _____ Model _____

Engine _____ AT MT CVT PSD (circle which applies)

Procedure

1. Turn the ignition on and note the warning lights on the instrument panel. Check for ABS, TC, ETC, VSC, or similar lights related to the antilock brake, traction, and stability control systems. Note which system lights are present.

   Lights _____

2. Use the service information to determine the type of system on this vehicle. Summarize your findings.

   _____

   _____

   _____

3. Use the service information to determine the location of traction and/or stability control system components. Note the locations. _____

   _____

   _____

   _____

Lab Worksheet 16-5

Name _____ Station _____ Date _____

Describe Regenerative Braking

ASE Education Foundation Correlation

This lab worksheet addresses the following **MLR** task:

5.G.2. Describe the operation of a regenerative braking system. **(P-3)**

Description of Vehicle

Year _____ Make _____ Model _____

Engine _____ AT MT CVT PSD (circle which applies)

Procedure

1. Define *regenerative braking*. _____

   _____

   _____

2. Describe what types of vehicles use regenerative braking. _____

   _____

   _____

3. What are the benefits of a vehicle having regenerative braking? _____

   _____

   _____

4. How do you think regenerative braking will affect the service life of the service brakes on a vehicle?

   _____

   _____

   _____

CHAPTER 17

Electrical/Electronic System Principles

Review Questions

1. Current trends have been to either enhance or replace _____ components with electrical components.

2. The term _____ applies to electrical circuits that contain bulbs, motors, and other devices.

3. The term _____ refers to integrated circuits and computers that are not directly tested.

4. Give a brief explanation of electricity in your own words. _____

   _____

   _____

5. Label the parts of the atom shown in Figure 17-1 (A through D).

   A. _____   B. _____

   C. _____   D. _____

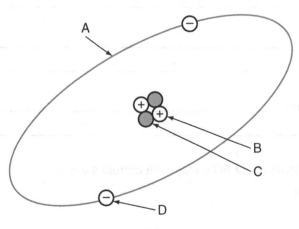

Figure 17-1

6. A copper atom, in its normal state with 29 protons and electrons, is called electrically _____.

7. _____ electricity is the term used to describe the accumulation of an electrical charge by a substance not normally conductive to electricity.

8. Define *voltage*. _____
_____

9. The amount of voltage being used by a component, wire, or connection is called _____.

10. Define *amperage*. _____
_____

11. Define *ohms*. _____
_____

12. The relationship commonly used to describe electrical behavior is called _____ law.

13. Place each letter in the correct location in Figure 17-2.

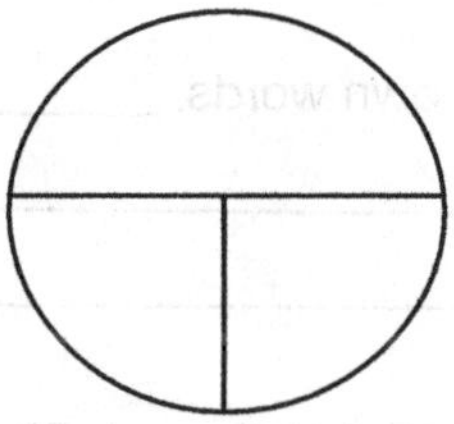

I
R
E

Figure 17-2

14. Explain what functions can and cannot be performed by volts, amps, and ohms using Ohm's law. _____
_____

15. Define *watts*. _____
_____

16. Describe alternating current. _____
_____

17. All of the following statements about AC voltage are correct except:
    a. AC has positive voltages.
    b. AC has zero voltage.
    c. AC changes between positive and negative voltage.
    d. AC eliminates negative voltages.

18. All of the following statements about DC voltage are correct except:
    a. DC alternates between positive and negative.
    b. DC is used by battery powered devices.
    c. DC voltages can vary.
    d. Some devices require AC to be turned into DC to operate.

19. Explain several advantages and disadvantages of using AC and DC. _____
    _____
    _____
    _____

20. *Circuit* is a term that refers to a complete electrical _____ from power to ground.

21. A _____ is a path for electrons to flow from a source of higher electrical potential to a source of lower potential.

22. List five components of a complete circuit.
    a. _____
    b. _____
    c. _____
    d. _____
    e. _____

23. Label the parts of the circuit shown in Figure 17-3 (A through F).
    A. _____   B. _____
    C. _____   D. _____
    E. _____   F. _____

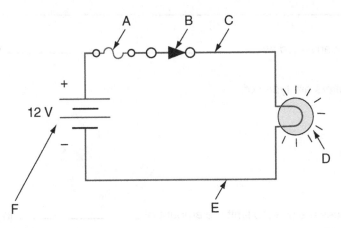

Figure 17-3

24. The battery is the source of high electrical _____ for the circuit.

25. Negative or ground refers to the _____ path of the circuit.

26. Anything that consumes electrical power is a _____.

27. _____ are the wires, terminals, and connectors that connect the parts of the circuit together.

28. Switches are types of circuit _____, used to turn the circuit on and off.

29. List three forms of circuit protection devices.
    a. _____
    b. _____
    c. _____

30. List the components of a basic circuit. _____
    _____

31. Define the principles of a series circuit. _____
    _____
    _____
    _____

32. Define the principles of a parallel circuit. _____
    _____
    _____
    _____
    _____

33. Automotive circuits are arranged as _____-_____ circuits.

34. Fuses and circuit breakers are types of:
    a. Control devices
    b. Loads
    c. Protection devices
    d. Conductors

35. Circuit protection devices are used to limit the amount of _____ in a circuit and prevent wire and component damage.

36. A vehicle has a blown 10 A fuse. *Technician A* says to replace the fuse with a higher amperage rated fuse to prevent the fuse from blowing again. *Technician B* says if the fuse continues to blow, a fault exists with the fuse. Who is correct?

    a. Technician A

    b. Technician B

    c. Both A and B

    d. Neither A nor B

37. Circuit breakers are often used in circuits with _____ because of sometimes sudden surges in current flow.

38. Multiple fuses are located together in a fuse box or _____ box.

39. A digital multimeter can be used to measure:

    a. amps

    b. ohms

    c. volts

    d. amps, ohms, and volts

40. *Technician A* says unwanted resistance in a circuit can cause a component, such as a light bulb, not to operate. *Technician B* says unwanted resistance in a circuit will cause the fuse to blow. Who is correct?

    a. Technician A

    b. Technician B

    c. Both A and B

    d. Neither A nor B

41. List four possible causes for a bulb not to light up.

    a. _____

    b. _____

    c. _____

    d. _____

42. A reading of 0 volts means that there is no difference in _____ between the two meter leads.

43. When referring to wire gauge, the larger the number of the gauge:

    a. The larger the wire diameter

    b. The smaller the wire diameter

    c. The more expensive the wire construction materials

    d. None of the above

44. A wiring _____ is a bundle of wires and connectors.

45. *Technician A* says terminals are the parts of a harness that connect two wires together. *Technician B* says terminals contain the connectors that connect two wires together. Who is correct?

    a. Technician A

    b. Technician B

    c. Both A and B

    d. Neither A nor B

46. Explain what is meant by a common ground. _____
_____
_____

47. Electrical current flow generates a _____ field around a wire.

48. Explain how electromagnetism is used in modern vehicles. _____
_____
_____
_____

49. A relay is used to:

    a. Control circuit current flow levels

    b. Control high current flow components

    c. Protect a circuit

    d. Control circuit current flow levels, control high current flow components, and protect a circuit

50. Explain the operation of a relay. _____
_____
_____
_____

51. Explain the basic operation of a DC motor. _____
_____
_____
_____

52. Describe how the AC generator produces electricity to power the electrical system.
_____
_____
_____
_____

53. The basis of all electronics is the _____.

54. All of the following are types of electronic devices except:
    a. Diodes
    b. LEDs
    c. Transformers
    d. Transistors

55. Explain what a diode is and its common uses in the automobile. _____
    _____
    _____
    _____

56. _____ are solid-state switches.

57. Explain what ESD is and how to prevent against accidental ESD damage.
    _____
    _____
    _____
    _____

Activities

1. Solve the series circuits in Figure 17-4 through Figure 17-8.

   a. Explain the relationship between current flow and resistance in the circuit. _____
   _____
   _____
   _____

   b. Describe the properties of series circuits. _____
   _____
   _____
   _____

   c. Why do you think series circuits are not generally used in automotive applications?
   _____
   _____
   _____
   _____

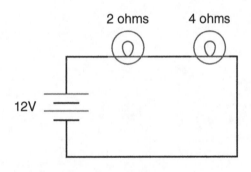

Resistance total =
Amperage total =
Voltage drop 1 =
Voltage drop 2 =

Figure 17-4

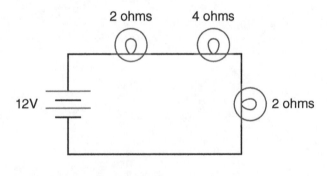

Resistance total =
Amperage total =
Voltage drop 1 =
Voltage drop 2 =
Voltage drop 3 =

Figure 17-5

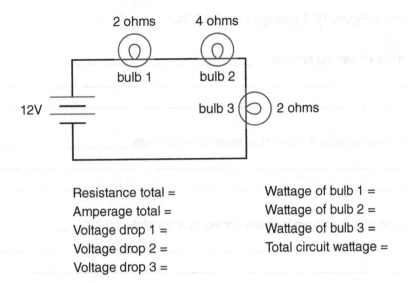

2 ohms   4 ohms

bulb 1   bulb 2

12V

bulb 3   2 ohms

Resistance total =   Wattage of bulb 1 =
Amperage total =   Wattage of bulb 2 =
Voltage drop 1 =   Wattage of bulb 3 =
Voltage drop 2 =   Total circuit wattage =
Voltage drop 3 =

Figure 17-6

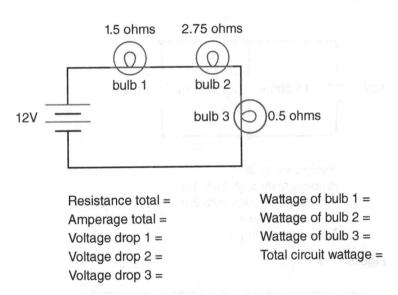

1.5 ohms   2.75 ohms

bulb 1   bulb 2

12V

bulb 3   0.5 ohms

Resistance total =   Wattage of bulb 1 =
Amperage total =   Wattage of bulb 2 =
Voltage drop 1 =   Wattage of bulb 3 =
Voltage drop 2 =   Total circuit wattage =
Voltage drop 3 =

Figure 17-7

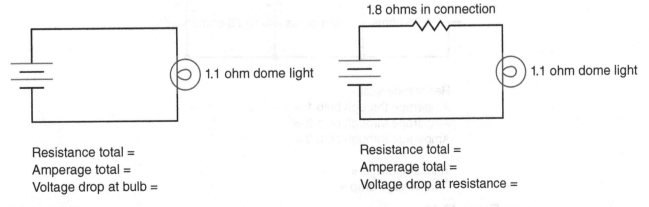

1.8 ohms in connection

1.1 ohm dome light   1.1 ohm dome light

Resistance total =   Resistance total =
Amperage total =   Amperage total =
Voltage drop at bulb =   Voltage drop at resistance =

Figure 17-8

2. Solve the parallel circuits in Figure 17-9 through Figure 17-11.

a. Describe the properties of parallel circuits. _____
_____
_____

b. Explain how parallel circuits differ in operation from series circuits. _____
_____
_____

c. What happens to current flow as branches are added to a parallel circuit? _____
_____
_____

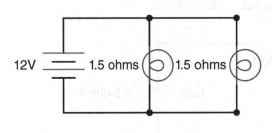

Resistance total =
Amperage through bulb 1 =
Amperage through bulb 2 =
Total amperage =
Total voltage drop =

Figure 17-9

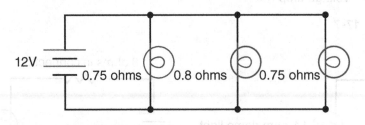

Resistance total =
Amperage through bulb 1 =
Amperage through bulb 2 =
Amperage through bulb 3 =

Total amperage =
Total voltage drop =

Figure 17-10

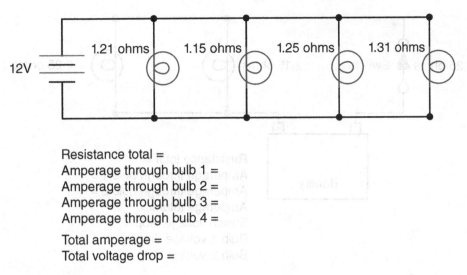

Resistance total =
Amperage through bulb 1 =
Amperage through bulb 2 =
Amperage through bulb 3 =
Amperage through bulb 4 =

Total amperage =
Total voltage drop =

Figure 17-11

3. Solve the series–parallel circuits in Figure 17-12 through Figure 17-14.

a. Describe how series–parallel circuits apply to automotive electrical systems.

_____

_____

_____

b. Why are most circuits of the series–parallel design? _____

_____

_____

_____

c. What are the advantages of series–parallel circuits compared to series circuits and parallel circuits?

_____

_____

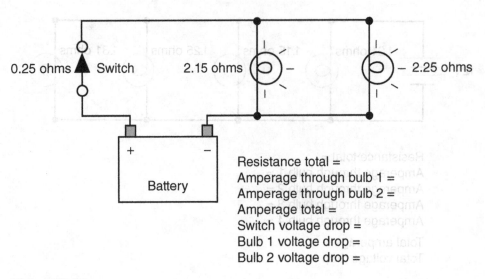

Resistance total =
Amperage through bulb 1 =
Amperage through bulb 2 =
Amperage total =
Switch voltage drop =
Bulb 1 voltage drop =
Bulb 2 voltage drop =

Figure 17-12

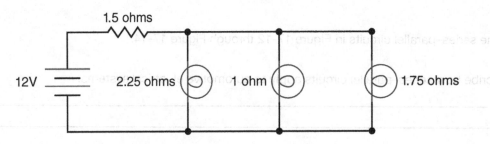

Resistance total =
Amperage total =
Volatge drop of unwanted resistance =
Voltage drop of light bulbs =

Figure 17-13

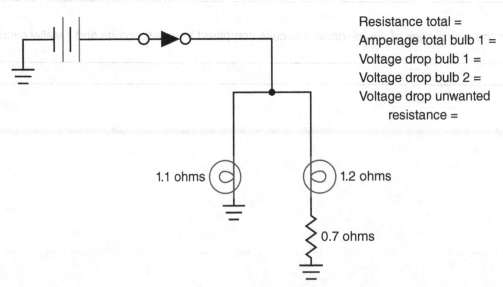

Resistance total =
Amperage total bulb 1 =
Voltage drop bulb 1 =
Voltage drop bulb 2 =
Voltage drop unwanted
  resistance =

Figure 17-14

4. Constructing an electromagnet

   A simple electromagnet can be built using a length of insulated wire and a nail or a screwdriver. Wrap a length of wire around a piece of metal, as shown in Figure 17-15. Connect the ends of the wire to a 9V battery. The magnetic field produced should be strong enough to pick up small metal objects, such as paper clips.

   a. List at least five examples of automotive components that use electromagnetism.

   _____
   _____
   _____
   _____
   _____

   b. What do you think would result from using more voltage to power the electromagnet?

   _____
   _____

   c. How would increasing the size of the wire and the number of loops formed by the wire affect electromagnet strength?

   _____
   _____

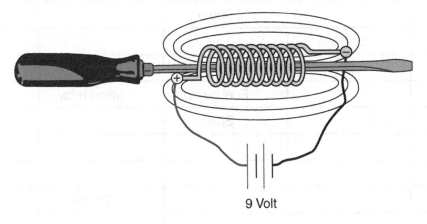

9 Volt

Figure 17-15

5. Label the types of circuit protection shown in Figure 17-16 (A through F).

   A. _____   B. _____
   C. _____   d. _____
   E. _____   F. _____

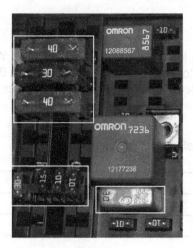

Figure 17-16

6. Refer to Figure 17-17 and Figure 17-18 to complete the following table:

TABLE 17–1   Refer to Figure 17–17 and Figure 17–18 and complete the following table.

Fuse/Relay Location #	Fuse Size	Circuit Description	Fuse/Relay Type	Fuse Color
38				
		Radio		
	40			
6				
		Fuel pump	Micro relay	
		Spare		
25				
				Orange
			Diode	
12				

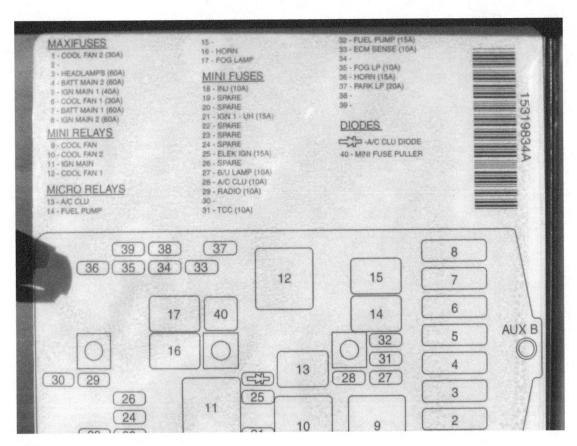

Figure 17-17

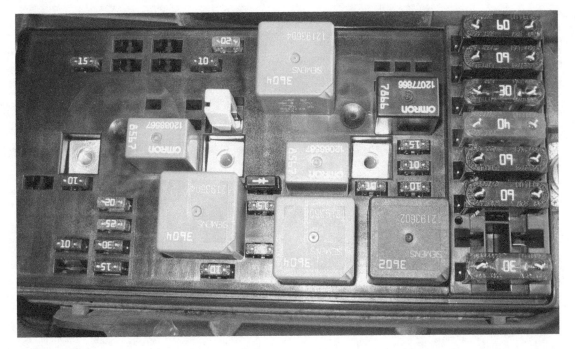

Figure 17-18

Lab Worksheet 17-1

Name _____ Station _____ Date _____

Knowledge of Ohm's Law: Series Circuits

ASE Education Foundation Correlation

This lab worksheet addresses the following **MLR** task:

6.A.2. Demonstrate knowledge of electrical/electronic series, parallel, and series-parallel circuits using principles of electricity (Ohm's Law). **(P-1)**

Procedure

Use your knowledge of Ohm's law to solve the following circuit problems.

1. Rt = _____   At = _____

    Vd _____   _____

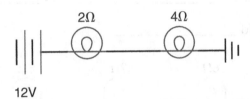

2. Rt = _____   At = _____

    Vd _____   Vd _____

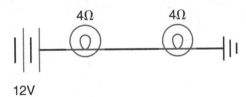

3. Rt = _____   At = _____

    Vd _____   Vd _____ _____

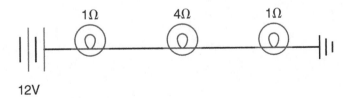

4. Rt = _____   At = _____

    Vd _____   Vd _____ _____

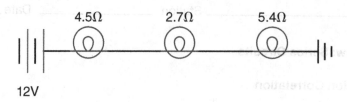

5. V = _____   At = _____

    Vd _____   Vd _____ _____

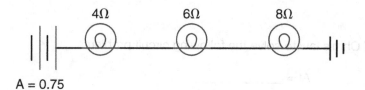

6. Rt = _____   At = _____   R2 = _____

    Vd _____   Vd _____ _____

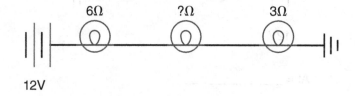

Lab Worksheet 17-2

Name _____ Station _____ Date _____

Knowledge of Ohm's Law: Parallel Circuits

ASE Education Foundation Correlation

This lab worksheet addresses the following **MLR** task:

6.A.2. Demonstrate knowledge of electrical/electronic series, parallel, and series-parallel circuits using principles of electricity (Ohm's Law). **(P-1)**

Procedure

Use your knowledge of Ohm's law to solve the following circuit problems.

1. Rt = _____   At = _____

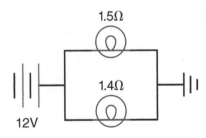

3Ω

3Ω

12V

2. Rt = _____   At = _____

1.5Ω

1.4Ω

12V

3. V = _____   Rt = _____

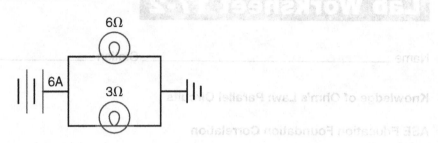

4. Rt = _____   At = _____

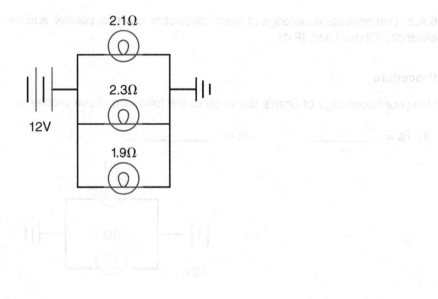

Lab Worksheet 17-3

Name _____ Station _____ Date _____

Knowledge of Ohm's Law: Series-Parallel Circuits

ASE Education Foundation Correlation

This lab worksheet addresses the following **MLR** task:

6.A.2. Demonstrate knowledge of electrical/electronic series, parallel, and series-parallel circuits using principles of electricity (Ohm's Law). **(P-1)**

Procedure

Use your knowledge of Ohm's law to solve the following circuit problems.

1. Rt = _____   At = _____

    Vd _____   Vd _____ Vd _____

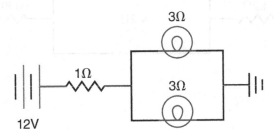

2. Rt = _____   At = _____

    Vd _____   Vd _____ Vd _____

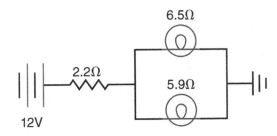

3. Rt = _____   At = _____

   Vd _____   Vd _____

   Vd _____   Vd _____

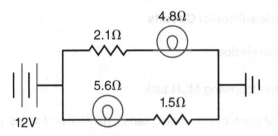

4. Rt = _____   At = _____   Vd _____   Vd _____

   Vd _____   Vd _____   Vd _____

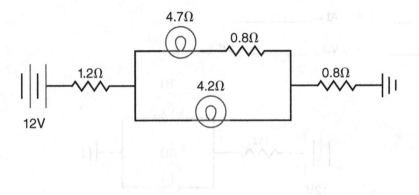

Lab Worksheet 17-4

Name _____ Station _____ Date _____

ID Components

ASE Education Foundation Correlation

This lab worksheet addresses the following **MLR** task:

6.A.11. Identify electrical/electronic system components and configuration. **(P-1)**

Description of Vehicle

Year _____ Make _____ Model _____

Engine _____ AT MT CVT PSD (circle which applies)

Procedure

Using the vehicle, owner's manual, and service information, locate the following components for this vehicle.

1. 12-volt battery _____   2. Auxiliary battery _____

3. Fuse/relay box 1 _____   4. Fuse/relay box 2 _____

5. Fuse/relay box 3 _____   6. Fuse/relay box 4 _____

If the vehicle is a hybrid-electric vehicle (HEV) or electric vehicle (EV), note the locations of the following items.

1. 12-volt battery _____   2. High-voltage battery _____

3. High-voltage disconnect _____   4. Inverter assembly _____

Locate the following:

1. Fuel pump relay/fuse _____

2. Body control module (BCM) _____

3. Remote control door lock receiver (RKE) _____

4. AC Generator output rating _____

5. Low beam headlight part number _____

CHAPTER 18

Basic Electrical/Electronic System Service

Review Questions

1. When working on the electrical system, _____ should be your first priority.

2. Electrical shocks can also cause _____ due to the heat caused by current flow.

3. High-voltage wiring found in hybrid vehicles is which color?
   a. Orange
   b. Yellow
   c. Red
   d. Blue

4. List six safety precautions for working around automotive batteries.
   a. _____
   b. _____
   c. _____
   d. _____
   e. _____
   f. _____

5. A digital multimeter (DMM) is the primary tool to make _____, _____, and resistance measurements.

6. A _____ can be used to check for power, ground, current flow, and continuity.

7. A current _____ is used to make inductive amperage measurements.

8. Describe electrostatic discharge and why it is dangerous for electronic components.

_____

_____

_____

_____

9. Voltage is electrical _____ and is measured as the difference in potential between _____ points.

10. List the steps to perform a voltage measurement. _____

_____

_____

_____

_____

11. Why do DMMs have high internal resistance? _____

_____

_____

_____

12. How would using a DMM with low internal resistance affect a circuit when used to test voltage?

_____

_____

_____

13. If using a DMM with manual voltage selections, which of the following settings would be used to measure vehicle battery voltage?
   a. 400-mv scale
   b. 4-volt scale
   c. 40-volt scale
   d. 400-volt scale

14. Define the term voltage drop. _____

_____

_____

15. *Technician A* says a voltage drop test can indicate excessive resistance in a circuit. *Technician B* says a voltage drop can be used to diagnose poor connections. Who is correct?

    a. Technician A

    b. Technician B

    c. Both A and B

    d. Neither A nor B

16. Voltage drop testing is a _____ test, meaning that the circuit must be in operation.

17. A voltage drop reading of 0.000 volts was obtained while testing a body ground. *Technician A* says the reading indicates that a very good ground condition is present. *Technician B* says the test may not have been performed correctly. Who is correct?

    a. Technician A

    b. Technician B

    c. Both A and B

    d. Neither A nor B

18. The small amount of movement between terminals that can lead to wear and a resistive connection is called:

    a. Thermal cycling

    b. Fretting

    c. Voltage drop

    d. High resistance

19. To measure the resistance of a component or circuit, the power must be _____ before taking the measurement.

20. On a DMM, the Ω symbol is used to designate:

    a. Current flow

    b. Voltage

    c. Resistance

    d. None of the above

21. Explain what is meant by continuity and why you would test it. _____

    _____

    _____

    _____

22. A resistance test is a _____ test, meaning the circuit is not operating.

23. Explain why a resistance test will not always find the cause of an excessive voltage drop or decreased current flow. _____

_____

_____

_____

24. Current flow in a circuit is directly _____ to the resistance of the circuit.

25. While discussing measuring amperage with a DMM, *Technician A* says incorrect DMM setup can cause incorrect amperage readings to be displayed by the meter. *Technician B* says incorrect DMM setup can cause the fuse in the meter to blow. Who is correct?

    a. Technician A

    b. Technician B

    c. Both A and B

    d. Neither A nor B

26. When testing current flow with a DMM, the meter must be placed:

    a. parallel to the circuit.

    b. in series with the circuit.

    c. in series–parallel with the circuit.

    d. none of the above.

27. To measure current flow above 10 amps, an _____ clamp is used.

28. _____ are used to represent components in electrical wiring diagrams.

29. In a wiring diagram, a component drawn in a dashed line indicates:

    a. the complete component is shown.

    b. not all of the component is shown.

    c. the component is not part of the circuit.

    d. the component is not accessible for testing.

30. Describe how color can be used to help the technician to understand wiring diagrams. _____

_____

_____

_____

31. Explain how to break a wiring diagram down into sections to understand circuit operation and how it can be tested. _____

_____

_____

_____

32. Explain the three types of circuit faults and how each affects circuit operation.

    _____

    _____

    _____

    _____

    _____

33. A vehicle has one headlight that is much brighter than the other headlight. Which type of circuit fault is most likely the cause?

    a. An open circuit

    b. A short to power

    c. A short to ground

    d. High resistance

34. An open circuit will cause a component to:

    a. draw excessive amperage

    b. draw excessive voltage

    c. draw no current

    d. have excessive resistance

35. A _____ circuit can be either to power or to ground.

36. Explain why a short to ground can be dangerous. _____

    _____

    _____

37. Before you begin to try to diagnose a short to ground, check to see if any aftermarket _____ have been recently installed in the vehicle.

38. Explain how to diagnose a short to ground. _____

    _____

    _____

    _____

39. High resistance in a circuit can cause all of the following except:

    a. Dim lights

    b. Excessive current flow

    c. Slow motor operation

    d. Inoperative components

40. Unwanted resistance can also cause connectors to _____ and melt.

41. Unwanted resistance in the circuit causes a _____-than-normal voltage drop on the actual load in the circuit.

42. A _____ pack or splice connector is used to connect several circuits together.

43. The first step when testing a fuse is _____ the fuse.

44. What will be required if a fuse keeps blowing? _____

_____

_____

_____

45. Describe how a circuit breaker functions. _____

_____

_____

46. Describe how to repair a section of broken wire. _____

_____

_____

_____

47. Do not force _____ into the terminals while you are performing tests as this can spread the terminal apart or cause other damage, resulting in open circuits or intermittent connections.

48. Relays typically have a coil resistance of about:
    a. 10–20 ohms
    b. 25–75 ohms
    c. 50–150 ohms
    d. 100–200 ohms

49. Explain how to load-test a relay. _____

_____

_____

_____

_____

Activities

1. Label the electrical tools shown in Figure 18-1 (a through i).

   a. _____   b. _____

   c. _____   d. _____

   e. _____   f. _____

Figure 18-1a

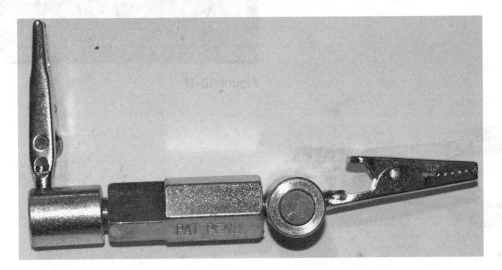

Figure 18-1b

Figure 18-1c

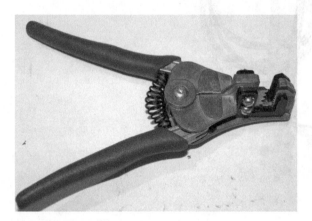

Figure 18-1d

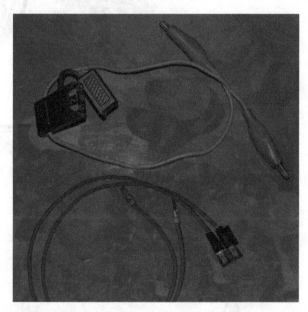

Figure 18-1f

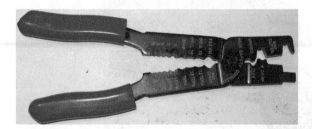

Figure 18-1e

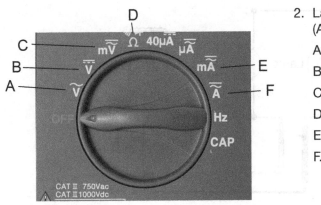

Figure 18-2

2. Label the functions of the meter shown in Figure 18-2 (A through F).

A. _____

B. _____

C. _____

D. _____

E. _____

F. _____

3. Label the types of tests shown in Figure 18-3 (a through c).

a. _____   b. _____

c. _____

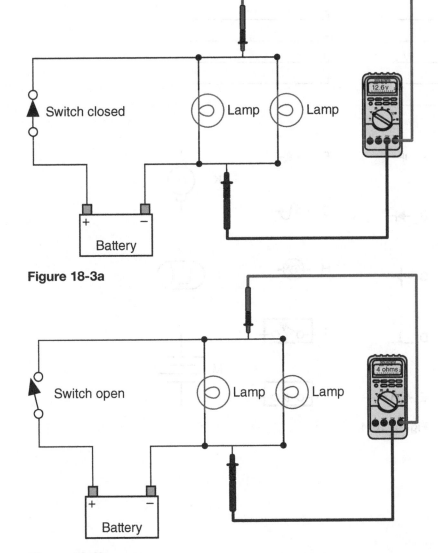

Figure 18-3a

Figure 18-3b

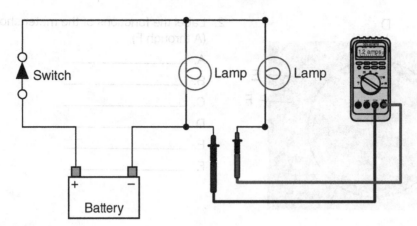

Figure 18-3c

4. Label the schematic symbols shown in Figure 18-4.

A. _____   B. _____

C. _____   D. _____

E. _____   F. _____

G. _____   H. _____

I. _____   J. _____

K. _____   L. _____

M. _____

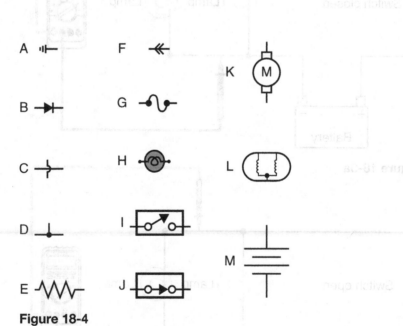

Figure 18-4

5. Label the terminals of the relays shown in Figure 18-5.

A. _____   B. _____

C. _____   D. _____

E. _____   F. _____

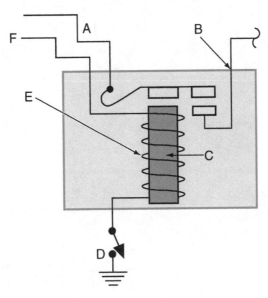

Figure 18-5

6. Match the following blade-type fuse ratings to the fuse colors.

3 amp	Green
5 amp	Tan/brown
10 amp	Yellow
15 amp	Violet
20 amp	Red
25 amp	Blue
30 amp	Clear

Lab Activity 18-1

Name _____ Date _____ Instructor _____

Using a DMM

Obtain from your instructor several automotive exterior lights with sockets. Bulb types 194, 1156, and 1157 (or similar) work well. Construct circuits using the diagrams below and record your findings.

1. **(Figure 18-6)** Series circuit.

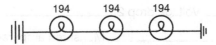

Figure 18-6 Series circuit.

Battery voltage: _____ Circuit resistance: _____

Voltage drop 1: _____ Voltage drop 2: _____ Voltage drop 3: _____

Ohm's law predicted amperage: _____ Measured amperage: _____

2. **(Figure 18-7)** Series circuit.

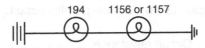

Figure 18-7 Series circuit

Battery voltage: _____ Circuit resistance: _____

Voltage drop 1: _____ Voltage drop 2: _____

Ohm's law predicted amperage: _____ Measured amperage: _____

3. **(Figure 18-8)** Parallel circuit.

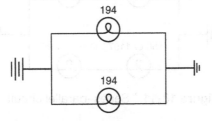

Figure 18-8 Parallel circuit.

Battery voltage: _____ Circuit resistance: _____

Voltage drop 1: _____ Voltage drop 2: _____

Ohm's law predicted amperage: _____ Measured amperage: _____

4. **(Figure 18-9)** Parallel circuit.

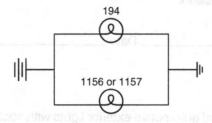

Figure 18-9 Parallel circuit.

Battery voltage: _____ Circuit resistance: _____

Voltage drop 1: _____ Voltage drop 2: _____

Ohm's law predicted amperage: _____ Measured amperage: _____

5. **(Figure 18-10)** Series–parallel circuit.

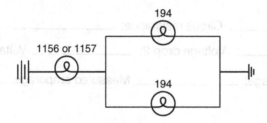

Figure 18-10 Series–parallel circuit.

Battery voltage: _____ Circuit resistance: _____

Voltage drop 1: _____ Voltage drop 2: _____

Voltage drop 3: _____ Voltage drop 4: _____

Ohm's law predicted amperage: _____ Measured amperage: _____

6. **(Figure 18-11)** Series–parallel circuit.

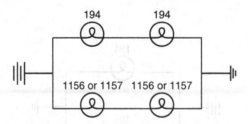

Figure 18-11 Series–parallel circuit.

Battery voltage: _____ Circuit resistance: _____

Voltage drop 1: _____ Voltage drop 2: _____ Voltage drop 3: _____

Voltage drop 4: _____ Voltage drop 5: _____ Voltage drop 6: _____

Ohm's law predicted amperage: _____ Measured amperage: _____

7. **(Figure 18-12)** Series–parallel circuit.

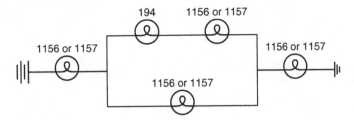

Figure 18-12 Series–parallel circuit.

Battery voltage: _____ Circuit resistance: _____ Voltage drop 1: _____

Voltage drop 2: _____ Voltage drop 3: _____ Voltage drop 4: _____

Voltage drop 5: _____ Voltage drop 6: _____ Voltage drop 7: _____

Ohm's law predicted amperage: _____ Measured amperage: _____

7. (Figure 16-12) Series-parallel circuit.

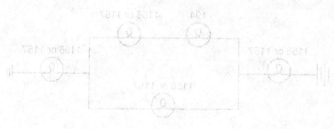

Figure 16-12   Series-parallel circuit.

Battery voltage: _____   Circuit resistance: _____   Voltage drop 1: _____

Voltage drop 2: _____   Voltage drop 3: _____   Voltage drop 4: _____

Voltage drop 5: _____   Voltage drop 6: _____   Voltage drop 7: _____

Ohm's law predicted amperage: _____   Measured amperage: _____

Lab Worksheet 18-1

Name _____ Station _____ Date _____

Wiring Diagrams

ASE Education Foundation Correlation

This lab worksheet addresses the following **MLR** task:

6.A.3. Use wiring diagrams to trace electrical/electronic circuits. **(P-1)**

Procedure

This wiring diagram represents a real vehicle you are testing. The circuit is for the radiator cooling fan with the fan operating normally. Based on the diagram and the operating condition above, record what a DMM would measure with the positive lead placed at the locations specified and the negative lead on battery ground.

The fan is running during this testing.

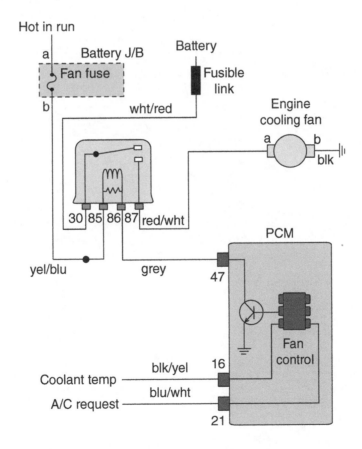

Test Location	Expected DMM Reading
1. Terminal b at cooling fan motor	
2. Terminal 86 at cooling fan relay	
3. Terminal 30 at cooling fan relay	
4. Terminal 87 at cooling fan relay	
5. Terminal b at battery J/B	
6. Terminal 86 at cooling fan relay	
7. Terminal 47 at PCM	

Lab Worksheet 18-2

Name _____ Station _____ Date _____

DMM Use

ASE Education Foundation Correlation

This lab worksheet addresses the following **MLR** task:

6.A.4. Demonstrate proper use of a digital multimeter (DMM) when measuring source voltage, voltage drop (including grounds), current flow, and resistance. **(P-1)**

Procedure

1. Draw and label the circuit you are testing in the space below.

2. Record the battery voltage _____

3. Measure and record total circuit resistance _____

4. Connect the circuit to power and ground. Record the voltage drop of each component.

   _____ Vdrop   _____ Vdrop   _____ Vdrop

   _____ Vdrop   _____ Vdrop   _____ Vdrop

   _____ Vdrop   _____ Vdrop   _____ Vdrop

5. Configure the DMM to measure amperage. Connect the DMM to the circuit and measure total circuit amperage.

   Amperage reading _____

6. Reset and turn off the DMM.

7. Describe the type of circuit you tested. _____

8. Explain how this circuit operates and the reading you obtained. _____

_____

_____

_____

Instructor's check _____

Lab Worksheet 18-3

Name _____ Station _____ Date _____

Circuit Faults

ASE Education Foundation Correlation

This lab worksheet addresses the following **MLR** task:

6.A.5. Demonstrate knowledge of the causes and effects from shorts, grounds, opens, and resistance problems in electrical/electronic circuits. **(P-1)**

Description of Vehicle

Year _____ Make _____ Model _____

Engine _____ AT MT CVT PSD (circle which applies)

Procedure

1. Explain an electrical short circuit. _____
   _____
   _____

2. What types of concerns can be caused by a short to power? _____
   _____
   _____

3. What types of concerns can be caused by a short to ground?   _____
   _____
   _____

4. Describe how to diagnose a short circuit. _____
   _____
   _____

5. Explain what is an open electrical circuit. _____
   _____
   _____

6. Explain what types of concerns are caused by open circuits. _____
_____
_____

7. Describe how to diagnose an open circuit. _____
_____
_____

8. Explain how unwanted resistance can affect a circuit. _____
_____
_____

9. Describe what types of concerns can be caused by unwanted resistance in a circuit. _____
_____
_____

10. Explain how to diagnose unwanted resistance in a circuit. _____
_____
_____

Lab Worksheet 18-4

Name _____ Station _____ Date _____

Test Light

ASE Education Foundation Correlation

This lab worksheet addresses the following **MLR** task:

6.A.6. Use a test light to check operation of electrical circuits. **(P-2)**

Description of Vehicle

Year _____ Make _____ Model _____

Engine _____ AT MT CVT PSD (circle which applies)

Procedure

1. Explain how a test light can be used to check for voltage, ground, and current flow. _____
   _____
   _____
   _____

2. To test for voltage, connect the test light wire lead to _____

3. Use the probe end of the test light to check for voltage at a fuse box. Check on both sides of a fuse. Does the test light illuminate at both test points?   Yes _____   No _____

4. What does it mean if a fuse only shows voltage on one side? _____
   _____

5. What does it mean if no voltage is present on either side of a fuse? _____
   _____

6. To check for ground, connect the test light wire lead to _____.

7. Use the probe end of the test light to check for ground at an under hood location.

   Instructor's check _____

8. Use the probe end of the test light to check for ground at an under dash location.

   Instructor's check _____

9. Why does the test light illuminate when the probe is touched to a ground location?

_____

_____

10. Describe how a test light can be connected to a circuit to show if current is flowing through the circuit.

_____

_____

11. Once finished, put away all tools and equipment and clean up the work area.

Instructor's check _____

Lab Worksheet 18-5

Name _____ Station _____ Date _____

Jumper Wires

ASE Education Foundation Correlation

This lab worksheet addresses the following **MLR** task:

6.A.7. Use fused jumper wires to check operation of electrical circuits. **(P-2)**

Description of Vehicle

Year _____ Make _____ Model _____

Engine _____ AT MT CVT PSD (circle which applies)

Procedure

1. Explain how fused jumper wires can be used electrical circuit faults. _____

   _____

   _____

   _____

2. Describe the safety procedures for working with fused jumper wires. _____

   _____

   _____

3. From a vehicle supplied by your instructor, locate and remove the horn relay from the fuse/relay box. Using a wiring diagram and a DMM, determine which connectors at the fuse/relay box are for the relay coil power, relay coil ground, power to the relay switch, and the circuit to the horn. Draw the configuration in the space below.

   Instructor's check _____

4. With the KOEO, connect a fused jumper wire between the terminal that supplies power to the relay switch and the wire to the horn. Did the horn sound?   Yes _____   No _____

5. If the horn sound, what does your test indicate? _____

   _____

6. If the horn did not sound, what could this indicate? _____

_____

7. Once finished, put away all tools and equipment and clean up the work area.

Instructor's check _____

Lab Worksheet 18-6

Name _____ Station _____ Date _____

Inspect Fuses

ASE Education Foundation Correlation

This lab worksheet addresses the following **MLR** task:

6.A.9. Inspect and test fusible links, circuit breakers, and fuses; determine necessary action. **(P-1)**

Description of Vehicle

Year _____ Make _____ Model _____

Engine _____ AT MT CVT PSD (circle which applies)

Procedure

1. Explain the purpose and operation of fusible links, circuit breakers, and fuses.

2. Visually inspect fusible link wire for damage. An open fusible link will often show signs of burning or the wire may pull apart easily if pulled. Note your findings. _____ _____

3. Many vehicles have bolt-in fusible links that look like large fuses. A visual inspection may show a burnt link, however, checking for power on both sides of the link is a better test. Use a test light or DMM to test for power on both terminals. Note your findings. _____ _____

4. Circuit breakers can be tested with an ohmmeter. Remove the circuit breaker and connect the DMM leads to the terminals. Record your reading.

   _____ Ohms

5. What does the reading on the DMM indicate about the circuit breaker? _____ _____

6. Fuses can be checked visually if removed from the fuse box or with a test light or DMM while installed. Describe how to test a fuse with a test light and with a DMM. _____ _____ _____

7. Once finished, put away all tools and equipment and clean up the work area.

   Instructor's check _____

Lab Worksheet 18-7

Name _____ Station _____ Date _____

Connectors and Terminals

ASE Education Foundation Correlation

This lab worksheet addresses the following **MLR** task:

6.A.10. Repair and/or replace connectors, terminal ends, and wiring of electrical/electronic systems (including solder repair). **(P-1)**

Description of Vehicle

Year _____ Make _____ Model _____

Engine _____ AT MT CVT PSD (circle which applies)

Procedure

Connector replacement

1. Begin by removing the damaged connector from the harness.

    Instructor's check _____

2. Compare the old and new connector to make sure the replacement connector is correct.

    Instructor's check _____

3. Using wire strippers, strip back approximately 10mm of insulation from the wires.

    Instructor's check _____

4. Slide a section of heat-shrink tubing over each wire before installing the terminals.

    Instructor's check _____

5. Install the replacement terminals over the ends of the wires and crimp the terminals in place with wire terminal pliers.

    Instructor's check _____

6. Make sure each wire terminal if solidly crimped to each wire by pulling on each terminal. If any terminal comes off a wire, install and crimp a new terminal onto the wire.

    Instructor's check _____

7. Once complete, clean up the work area.

    Instructor's check _____

Lab Worksheet 18-8

Name _____ Station _____ Date _____

Solder Repair

ASE Education Foundation Correlation

This lab worksheet addresses the following **MLR** task:

6.A.10. Repair and/or replace connectors, terminal ends, and wiring of electrical/electronic systems (including solder repair). **(P-1)**

Description of Vehicle

Year _____ Make _____ Model _____

Engine _____ AT MT CVT PSD (circle which applies)

Procedure

1. Begin by removing approximately one-half inch of insulation from the ends of the two wires to be soldered. Describe what tool you used to remove the insulation and why it is important to not damage the wire when removing the insulation. _____

   _____

   _____

2. Insert the exposed wire end into the wire splice connector and crimp the connector onto the wire. Explain why it is important for the wire to be secured properly into the connector before soldering. _____

   _____

   _____

3. Position the end of a length of solder where the wire enters the connector. Using the soldering gun, heat the connector (*not* the solder) until the solder begins to flow into the connector. Remove the soldering gun from the connector and let both cool.

   Instructor's check _____

4. Why is it important to heat the connector to the point that the solder melts and flows on its own? _____

   _____

5. Once the connection is soldered properly, what steps need to be taken to finish the wire repair and ensure it will be protected from corrosion and damage? _____

_____

6. Once complete, clean up the work area.

Instructor's check _____

Battery, Starting and Charging System Principles

Review Questions

1. The parts of a battery are the _____, the _____, and the electrolyte.

2. The electrolyte in an automotive battery is made of _____ acid and _____.

3. A fully charged automotive battery produces how much voltage per cell?
   a. 2 volts
   b. 2.1 volts
   c. 2.2 volts
   d. 12.6 volts

4. As an automotive battery discharges, the electrolyte becomes mostly _____.

5. The main function of the automotive battery is to power the _____ motor.

6. List and explain the four stages of battery operation. _____
   _____
   _____
   _____
   _____

7. Over time, the discharge and recharge process causes the battery to lose _____.

8. Which of the following affect how well a battery accepts a charge?
   a. Battery temperature
   b. Battery state of charge
   c. Plate sulfation
   d. All of the above

9. *Technician A* says overcharging a battery can shorten the battery life. *Technician B* says excessive vibration can shorten battery life. Who is correct?

   a. Technician A

   b. Technician B

   c. Both A and B

   d. Neither A nor B

10. Which type of automotive battery may require water to be added occasionally?

   a. Maintenance-free battery

   b. AGM battery

   c. Low-maintenance battery

   d. Deep-cycle battery

11. All of the following statements about hybrid vehicles are correct except:

   a. Some hybrid vehicles have both high-voltage and low-voltage batteries.

   b. The high-voltage battery is a different design than a lead-acid battery.

   c. The high-voltage battery is not easily accessible.

   d. A high-voltage system disconnect is used to isolate the high-voltage battery.

12. The high-voltage wiring found on hybrids is covered in bright _____ conduit.

13. The cold cranking amps rating is based on battery power available at _____°F.

14. The cranking amps rating is based on battery power available at _____°F.

15. The battery reserve capacity rating is used to rate in _____ battery output if the charging system fails.

16. Which battery rating is used to define a battery's size and terminal type?

   a. CCA

   b. AH

   c. CA

   d. BCI

17. The BCI group rating defines all of the following battery characteristics except:

   a. Terminal position

   b. Battery height

   c. CCA rating

   d. Overall battery size

18. List eight examples of battery handling and service safety precautions. _____
_____
_____
_____
_____
_____
_____

19. The function of the _____ is to spin the engine fast enough that the air-fuel mixture can be compressed and ignited so that combustion can begin.

20. When current flows through a conductor, a _____ field is produced around the conductor.

21. The parts of the electric motor responsible for switching the polarity of the magnetic fields are the:
    a. Armature and permanent magnets
    b. Permanent magnets and commutator
    c. Armature and brushes
    d. Brushes and commutator

22. Which of the following is the rotating part of the starter motor?
    a. Brushes
    b. Permanent magnets
    c. Armature
    d. Field coils

23. The solenoid completes the circuit to the starter motor and is used to pull a _____ that kicks out the drive gear into the flywheel.

24. Label the starter connections shown in Figure 19-1.
    a.
    b.
    c.

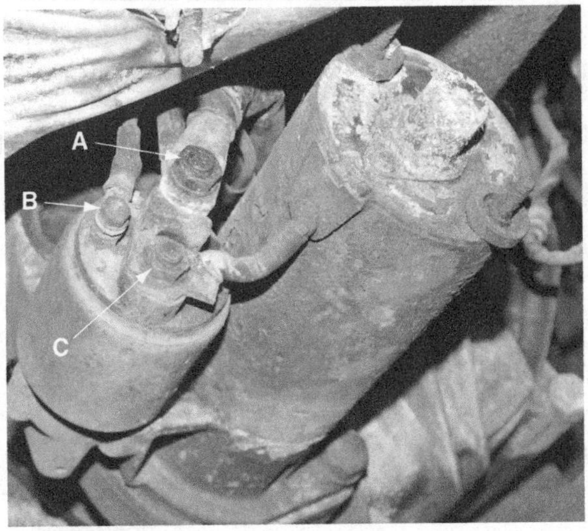

Figure 19-1

25. List the components of a typical starting system. _____
    _____
    _____

26. The three starter circuits are the _____ power, _____ ground, and the _____ circuit.

27. Explain the purpose of the park/neutral switch in the starting system. _____
    _____
    _____

28. Some antitheft systems use a starter _____ relay to prevent or allow starter operation based on the theft system status.

29. In newer vehicles, the ignition switch often does not supply power directly to the _____ circuit; instead, the switch is used as part of the control circuit that includes starter relays and _____ modules.

30. Vehicles that have Push to Start also use a _____ key to validate the starting process.

31. *Technician A* says full hybrids use a conventional 12-volt starter motor to crank the engine. *Technician B* says full hybrids use a high-voltage motor/generator to crank the engine. Who is correct?
    a. Technician A
    b. Technician B
    c. Both A and B
    d. Neither A nor B

32. True or False: Some nonhybrid vehicles use an engine idle stop–start function to improve fuel economy.
    _____

33. Once the engine has started, the _____ takes over powering the vehicle's electrical system.

34. The AC _____ is used to recharge the battery once the engine is started.

35. A modern generator produces as much as:
    a. 30 amps
    b. 50 amps
    c. 80 amps
    d. 140 amps

36. List the components of the generator rotor. _____
    _____
    _____

37. What is the purpose of the AOD or overrunning clutch? _____

_____

_____

38. The stator typically has _____ windings of wire.

39. _____ are used in the generator to convert AC into DC.

40. Which component controls the flow of current to the rotor coil?

    a. Stator

    b. Slip rings

    c. Voltage regulator

    d. Pole pieces

41. Explain the four factors of induction. _____

_____

_____

_____

_____

42. *Technician A* says generator field control is always on the ground side of the circuit. *Technician B* says voltage regulators may control the power or the ground side of the field. Who is correct?

    a. Technician A

    b. Technician B

    c. Both A and B

    d. Neither A nor B

43. Voltage regulators may be located:

    a. Internal to the generator

    b. External to the generator

    c. In the PCM

    d. All of the above

44. Describe how the PCM can be used to control generator output. _____

_____

_____

_____

Activities

1. Label the parts of the battery shown in Figure 19-2.

A. _____

B. _____

C. _____

D. _____

E. _____

F. _____

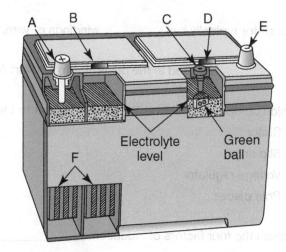

Electrolyte level

Green ball

Figure 19-2

2. Label the parts of the starters in Figure 19-3 (a through i).

A. _____

B. _____

C. _____

D. _____

E. _____

F. _____

G. _____

H. _____

I. _____

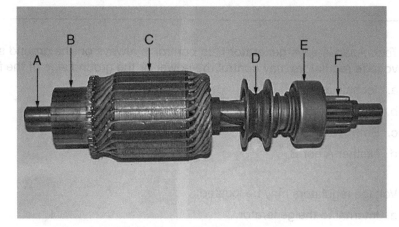

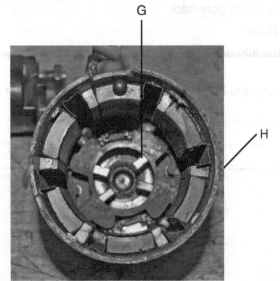

3. Obtain from your instructor two transmission pan type magnets, a two- to three-foot length of insulated single-strand wire like that used for armature windings on a fan motor, paper clips, or cotter pins, a Styrofoam cup or small box, and a 9V battery. Follow these steps to build a simple DC electric motor:

   • Strip the ends of the wire to remove about one-half of the insulation along the length of the last two inches of the wire. Each end should have bare copper exposed around half of the wire's circumference and insulation around the other half. This will allow an on/off connection through the wire.

   • Form the wire (coil) into a tightly wound loop, approximately 1 inch in diameter, leaving the two ends protruding out from the middle, as shown in Figure 19-4.

   • Mount the paper clips or cotter pins into the cup or box so that they can form a holding fixture for the wire.

   • Place the magnets so that they hold each other to the top section of the cup or box between the paper clips or cotter pins.

   • Place the coil into the paper clips or cotter pins. Adjust the height so that the coil can rotate very close to the magnet without touching.

   • Attach the 9V battery so that the positive is connected to one paper clip or cotter pin and the negative is connected to the other paper clip or cotter pin, as shown in Figure 19-5.

   • Some adjusting may be required, but once a good connection is made, the coil should easily spin on its own.

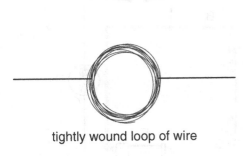

tightly wound loop of wire

Figure 19-4

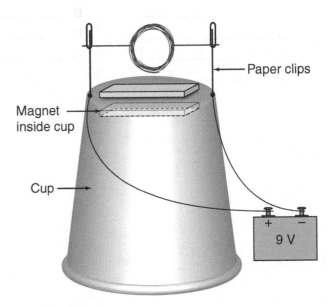

Paper clips

Magnet inside cup

Cup

9 V

Figure 19-5

a. Note the coil's direction of rotation. _____

b. Reverse the battery polarity and note the motor rotation. _____

c. Why did the motor respond as it did to the battery voltage change in polarity? _____

d. List the electrical motors used in modern automobiles that operate so that they reverse their direction of rotation. _____

e. List the electrical motors used in modern automobiles that rotate in only one direction. _____
_____

f. What would be the result if a motor, such as that for the cooling fan which only rotates one way, was hooked to the battery in reverse polarity? _____

g. Match the motor components with their electrical functions.

Paperclip/cotter pin	Armature
Wire	Brush
Magnets	Commutator
Wire insulation	Field coils

h. Explain why the wire's insulation is not completely removed. _____

i. What effect would using a smaller battery, such as a D cell, have on the motor and why? _____
_____

j. What effect would using more magnets have on the motor and why? _____

k. Why do you think the electric motors used in automobiles have more commutator segments than the simple motor you built? _____

l. What are the wear points on a DC motor? _____

m. What are some examples of problems that can affect the operation of a DC motor? _____
_____

4. Label the parts of the starting circuit shown in Figure 19-6.

A. _____   B. _____

C. _____   D. _____

E. _____   G. _____

G. _____

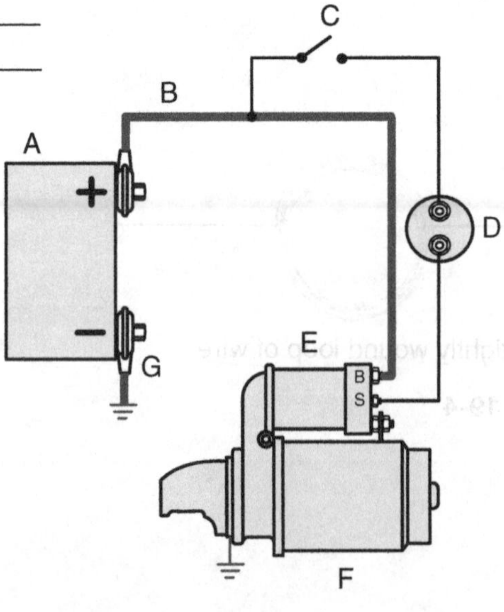

Figure 19-6

5. Label the parts of the charging system shown in Figure 19-7.

A. _____   B. _____

C. _____   D. _____

E. _____   F. _____

G. _____

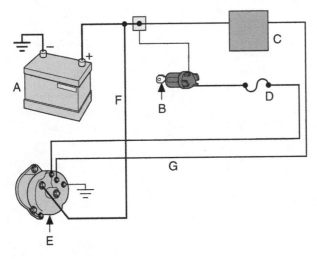

Figure 19-7

5. Label the parts of the charging system shown in Figure 19-7.

A. _____   B. _____

C. _____   D. _____

E. _____   F. _____

G. _____

Figure 19-7

Lab Activity 19-1

Name _____ Date _____ Instructor _____

Year _____ Make _____ Model _____

Identify Starting and Charging Components

1. Inspect the battery and note the following:

    CCA rating _____   CA rating _____

    RC rating _____   AH rating _____

    BCI group number _____

2. Note the location of the battery. _____

3. Locate the starter motor and note its location. _____

4. Using service information, determine the starter type. (Circle the type used)

    Direct drive _____ Gear reduction

5. Locate the AC generator and note its location. _____

6. Look on the generator or in the service information for the generator's rated output.

    Amps _____   Volts _____

7. Using the service information, determine how the generator field is controlled.

    _____

    _____

Lab Worksheet 19-1

Name _____ Station _____ Date _____

Research and TSB

ASE Education Foundation Correlation

This lab worksheet addresses the following **MLR** task:

6.A.1. Research applicable vehicle and service information, vehicle service history, service precautions, and technical service bulletins. **(P-1)**

Description of Vehicle

Year _____ Make _____ Model _____

Engine _____ AT MT CVT PSD (circle which applies)

Procedure

1. Inspect the battery and note the following:

   CCA rating _____   CA rating _____

   RC rating _____   A-h rating _____

   BCI group number _____

2. Note the location of the battery. _____

3. Locate the starter motor and note its location. _____

4. Using service information, determine the starter type.

   Direct drive _____   Gear reduction _____

5. Locate the AC generator and note its location _____

6. Look on the generator or in the service information for the generator's rated output.

   _____ amps   _____ volts

7. Using the service information, locate a technical service bulletin (TSB) for the starting or charging system.

   TSB number, date, and topic: _____

   _____

   Instructor's check _____

Lab Worksheet 19-2

Name _____ Station _____ Date _____

Load Test

ASE Education Foundation Correlation

This lab worksheet addresses the following **MLR** task:

6.B.2. Confirm proper battery capacity for vehicle application; perform battery capacity and load test; determine necessary action. **(P-1)**

Description of Vehicle

Year _____ Make _____ Model _____

Engine _____ AT MT CVT PSD (circle which applies)

Procedure

1. Locate the vehicle manufacturer's battery specifications.

   Recommended CCA _____   Recommended A-h rating _____

   Recommended RC rating _____   Other rating _____

2. Inspect the battery in the vehicle supplied by your instructor. Note any ratings on the battery decal.

   CA/CCA _____   A-h rating _____

   RC rating _____   Other _____

3. Based on your inspection, does the installed battery meet the manufacturer's specifications?

   Yes _____   No _____

4. Describe what concerns may be caused by a battery that has an insufficient CCA rating?

   _____

   _____

5. Describe how to perform a battery capacity or load test. _____

   _____

   _____

6. Connect the test equipment and determine the load to be placed on the battery.

   Load test amount _____

   Instructor's check _____

7. Perform the load test. Note your results. _____

   _____

   Instructor's check _____

8. Based on the load test, what is the condition of the battery? _____

9. Disconnect the tester and clean up the work area.

   Instructor's check _____

Lab Worksheet 19-3

Name _____ Station _____ Date _____

Conductance Test

ASE Education Foundation Correlation

This lab worksheet addresses the following **MLR** task:

6.B.2. Confirm proper battery capacity for vehicle application; perform battery capacity and load test; determine necessary action. **(P-1)**

Description of Vehicle

Year _____ Make _____ Model _____

Engine _____ AT MT CVT PSD (circle which applies)

Battery group number _____ CCA/CA _____

Procedure

1. Explain the battery conductance test. _____

_____

2. Tester used _____

3. Before testing, use a DMM to determine the battery state of charge.

   a. Open circuit voltage reading _____

   b. Is the battery sufficiently charged to continue?   Yes _____   No _____

4. Connect the conductance tester and program as necessary.

5. Perform the test and record the results. _____

_____

6. Based on this test, what is the condition of the battery? _____

_____

7. Remove any tools and equipment and clean up the work area.

Instructor's check _____

Lab Worksheet 19-4

Name _____ Station _____ Date _____

ID High-Voltage

ASE Education Foundation Correlation

This lab worksheet addresses the following **MLR** task:

6.B.7. Identify safety precautions for high voltage systems on electric, hybrid-electric, and diesel vehicles. **(P-2)**

Description of Vehicle

Year _____ Make _____ Model _____

Engine _____ AT MT CVT PSD (circle which applies)

Procedure

1. Using the service information, locate and summarize any safety precautions for working on or near the hybrid or electric drive system for this vehicle. _____

   _____

   _____

2. Using the service information, determine how to identify the high-voltage components for this vehicle.

   _____

   _____

3. What color is the high-voltage wiring in this vehicle? _____

4. Summarize the procedures to safely disconnect the high-voltage system on this vehicle.

   _____

   _____

   _____

5. What special precautions are necessary when working on the electrical system of a diesel-powered vehicle?

   _____

   _____

CHAPTER 20

Starting and Charging System Service

Review Questions

1. Describe four tools used in diagnosing the starting and charging systems.

   _____

   _____

   _____

   _____

2. Describe three tools used for servicing batteries. _____

   _____

   _____

   _____

3. List six safety precautions for working around batteries.

   a. _____

   b. _____

   c. _____

   d. _____

   e. _____

   f. _____

4. A starter should never be cranked for more than _____ seconds at a time without at least a _____ minute cool-off period.

5. HEV high-voltage wiring and components are identified by the _____ colored conduit and connections.

6. *Technician A* says all modern vehicles place the 12 V battery under the hood. *Technician B* says some 12 V batteries are located inside the passenger compartment. Who is correct?

    a. Technician A

    b. Technician B

    c. Both A and B

    d. Neither A nor B

7. Battery corrosion can be neutralized by using _____ and water as a paste.

8. List five types of problems the battery case should be inspected for. _____

    _____

    _____

    _____

    _____

9. *Technician A* says all lead-acid batteries have removable cell caps so that the electrolyte levels can be checked. *Technician B* says only maintenance-free batteries have removable cell caps. Who is correct?

    a. Technician A

    b. Technician B

    c. Both A and B

    d. Neither A nor B

10. If the terminal is loose on the post, an _____ connection problem can occur.

11. Side-post batteries can be _____ by overtightening the side-post connections.

12. List three things battery cables should be inspected for.

    a. _____

    b. _____

    c. _____

13. A loose battery from a missing _____ can be damaged by bouncing around.

14. Explain why it is important to use a memory-saving device when disconnecting a battery for service.

    _____

    _____

15. Many batteries have a built-in _____ to indicate the battery state of charge.

16. Explain what is meant by a battery voltage leak and how to test for it._____
_____
_____
_____

17. A battery is low on electrolyte: *Technician A* says the battery can be topped off with tap water. *Technician B* says the generator may be undercharging the battery, causing the low acid level. Who is correct?
    a. Technician A
    b. Technician B
    c. Both A and B
    d. Neither A nor B

18. A fully charged battery should have at least _____ volts.
    a. 12.6
    b. 12.2
    c. 12.9
    d. 12.0

19. While discussing battery charging: *Technician A* says the best way to recharge a battery is with a high charge rate over a long period of time. *Technician B* says a low charge rate over several hours is less likely to damage the battery. Who is correct?
    a. Technician A
    b. Technician B
    c. Both A and B
    d. Neither A nor B

20. List the steps to properly connect and set a battery charger. _____
_____
_____
_____

21. Which type of automotive battery requires special charging procedures?
    a. Maintenance-free batteries
    b. AGM batteries
    c. Low-maintenance batteries
    d. None of the above

22. Describe how to perform a battery load test. _____
_____
_____
_____

23. If the battery load capacity is unknown, explain another method to determine how much load to apply during a battery load test. _____

    _____

    _____

24. A battery load test is performed for _____ seconds.

25. A key-off drain, also called a _____ load, continues to draw power from the battery after the vehicle is shut off.

26. Which of the following methods can be used to test key-off battery drain?

    a. Inductive current clamp

    b. Test light in series with the battery and battery cable

    c. DMM in series with the battery and battery cable

    d. All of the above

27. Key-off battery drain should generally be less than:

    a. 200mA

    b. 100mA

    c. 50mA

    d. 10mA

28. List the proper order in which to connect two batteries for jump starting. _____

    _____

    _____

29. Describe the procedure to disable the high-voltage system on a hybrid vehicle.

    _____

    _____

    _____

    _____

30. Describe how to voltage drop test the starter control circuit. _____

    _____

    _____

    _____

31. The starting system in Figure 20-1 is being tested for a no-crank condition. With the key in the start position, the voltage reading shown in Figure 20-1 is observed. *Technician A* says this indicates that the neutral safety switch is open. *Technician B* says the solenoid may be faulty. Who is correct?

    a. Technician A

    b. Technician B

    c. Both A and B

    d. Neither A nor B

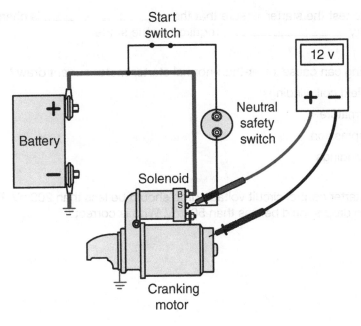

Figure 20-1

32. The circuit shown in Figure 20-2 is being tested for a no-crank condition. Based on the voltage reading shown, which is the mostly likely cause?

a. Faulty starter motor

b. Faulty neutral safety switch

c. Faulty ignition switch

d. Discharged battery

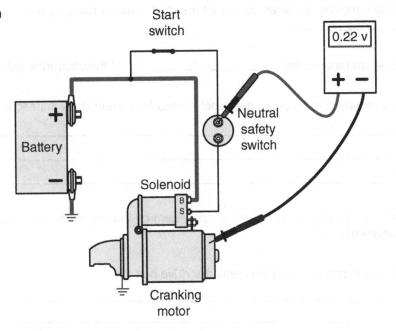

Figure 20-2

33. A vehicle's starter clicks rapidly when the key is turned to the start position: which is the most likely cause?

a. Low battery state of charge

b. Defective solenoid

c. Shorted starter windings

d. Defective park-neutral switch

34. Before attempting to test the starter, ensure that the _____ is charged and capable of producing the necessary _____ required by the starter.

35. Which of the following can cause lower-than-normal starter motor current draw?

    a. Shorted starter field coil windings

    b. Binding of the armature

    c. Low engine compression

    d. Open solenoid windings

36. *Technician A* says starter control circuit voltage drop should be less than 200mV. *Technician B* says starter motor circuit voltage drop should be less than 500mV. Who is correct?

    a. Technician A

    b. Technician B

    c. Both A and B

    d. Neither A nor B

37. If the voltage drops of the battery cables are excessive, remove and clean the battery cable connections at the _____ and the _____.

38. Before removing the starter motor from the vehicle, first disconnect the _____.

39. When removing a starter, do not let the starter motor hang by the _____ or control circuit _____.

40. Some starters require a _____ to set the clearance between the starter and the flywheel.

41. Explain why some starters are bench tested for current draw instead of being tested on the vehicle.

    _____

    _____

    _____

42. Begin the inspection of the charging system by looking at the _____ indicator light on the instrument panel.

43. List five things to check the generator drive belt for.

    a. _____

    b. _____

    c. _____

    d. _____

    e. _____

44. A loose generator drive belt can cause:

    a. An overcharge condition

    b. An undercharge condition

    c. A discharged battery

    d. An overcharge condition and an undercharge condition

45. A _____ battery can affect the generator's performance.

46. *Technician A* says for the charging system to operate properly, the battery must be in good condition. *Technician B* says battery condition should not affect generator operation. Who is correct?

    a. Technician A

    b. Technician B

    c. Both A and B

    d. Neither A nor B

47. A properly operating charging system should be able to produce at least _____ of the generator's rated output.

    a. 50%

    b. 60%

    c. 80%

    d. 90%

48. Describe how to test for AC voltage leaking from the generator. _____
    _____
    _____
    _____

49. List five problems that can cause an undercharge condition.

    a. _____

    b. _____

    c. _____

    d. _____

    e. _____

50. An open field winding will cause which of the following concerns?

    a. Overcharging

    b. Undercharging

    c. No charging

    d. None of the above

51. When testing generator output, you may need a _____ to check for DTCs and data for the charging system.

52. An _____ problem occurs when the field control circuit does not limit the amount of current that is supplied to the field.

53. When removing a generator, first disconnect the _____ to prevent damage to the electrical system.

Activities

1. A vehicle has a no-crank condition: When the headlights are turned on and the ignition turned to start, no noise is heard and the lights do not dim. Which of the following should be checked first and why?

    Battery connections   Ignition switch

    Battery voltage   Drive belt

    Starter connections   Generator connections

    Starter solenoid   Starter motor

2. Place the connections shown in Figure 20-3 in the correct order to jump-start a vehicle.

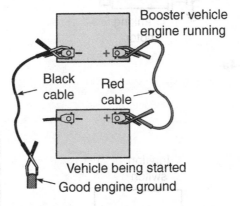

Booster vehicle engine running

Black cable   Red cable

Vehicle being started
Good engine ground

Figure 20-3

3. Identify the tests shown in Figure 20-4 (a through d).

    a. _____

    b. _____

    c. _____

    d. _____

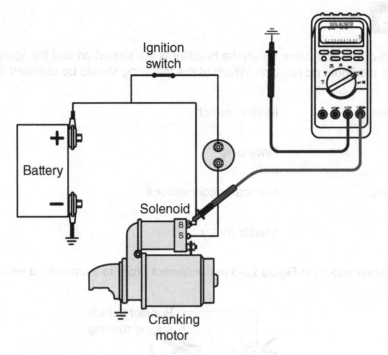

Figure 20-4a

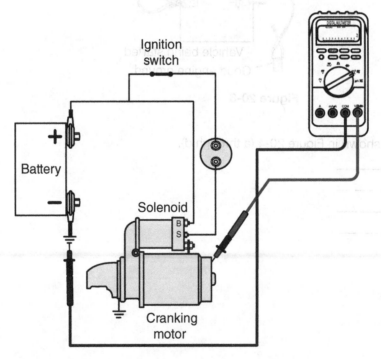

Figure 20-4b

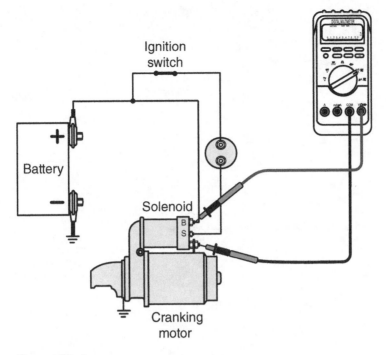

Ignition
switch

Battery

Solenoid

B
S

Cranking
motor

Figure 20-4c

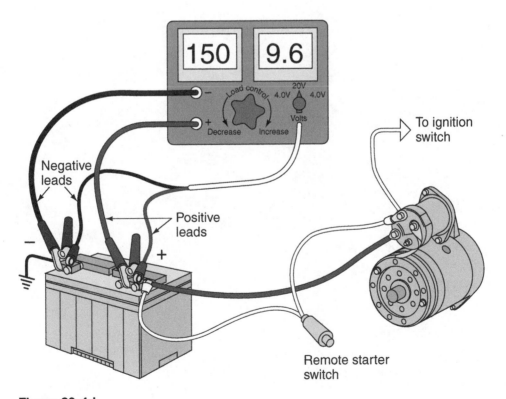

150   9.6

Load control

20V

4.0V   4.0V

Volts

Decrease   Increase

To ignition
switch

Negative
leads

Positive
leads

−   +

Remote starter
switch

Figure 20-4d

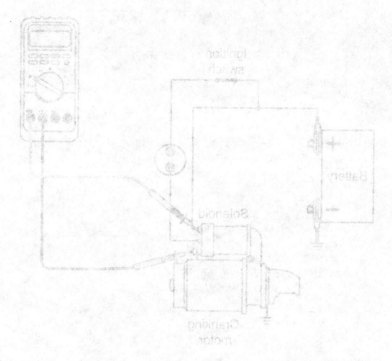

Ignition
switch

Solenoid

Battery

Cranking
motor

Figure 20-46

To ignition
switch

9.6

150

Negative
leads

Positive
leads

Remote starter
switch

Figure 20-54

Lab Worksheet 20-1

Name _____ Station _____ Date _____

ID HEV

ASE Education Foundation Correlation

This lab worksheet addresses the following **MLR** task:

1.A.7. Identify service precautions related to service of internal combustion engine of a hybrid vehicle. **(P-2)**

Description of Vehicle

Year _____ Make _____ Model _____

Engine _____ AT MT CVT PSD (circle which applies)

Procedure

⚠ **WARNING:** HIGH VOLTAGE VEHICLE ⚠
TO REDUCE THE RISK OF POSSIBLE SERIOUS INJURY (SHOCK OR BURN) OR DEATH:
COMPONENTS MARKED WITH THE HIGH VOLTAGE SYMBOL ⚠ CONTAIN HIGH VOLTAGE AND
HIGH TEMPERATURES AND SHOULD BE AVOIDED. 📖
SERVICE MUST BE PERFORMANCE BY QUALIFIED PERSONNEL ONLY.

1. Using the service information, locate and summarize any safety precautions for working on or near the engine or hybrid drive system for this vehicle. _____

   _____

2. Using the service information, locate and summarize any service precautions for working on the engine for this vehicle. _____

   _____

   _____

3. What color is the high-voltage wiring in this vehicle? _____

4. Summarize the procedures to safely disconnect the high-voltage system on this vehicle.

   _____

   _____

Lab Worksheet 20-2

Name _____ Station _____ Date _____

Parasitic Draw

ASE Education Foundation Correlation

This lab worksheet addresses the following **MLR** task:

6.A.8. Measure key-off battery drain (parasitic draw). **(P-1)**

Description of Vehicle

Year _____ Make _____ Model _____

Engine _____ AT MT CVT PSD (circle which applies)

Battery group number _____ CCA/CA _____

Procedure

1. Explain why a parasitic draw test is performed. _____
   _____

2. Check which of the three methods of testing parasitic draw will be used.
   _____ Test light
   _____ DMM
   _____ Inductive ammeter

3. Explain why using a test light is the least accurate parasitic draw test method. _____
   _____

4. Explain why disconnecting the battery to perform this test is not always a good decision. _____
   _____
   _____

5. Connect the test equipment and measure the parasitic draw.

   _____ amps

6. Locate the manufacturer's specification for allowable parasitic draw and record.

   _____ amps

7. Does the measured draw exceed the spec?   Yes _____   No _____

8. If the draw is excessive, what are the possible causes of the draw? _____

   _____

9. Remove any tools and equipment and clean up the work area.

                              Instructor's check _____

Lab Worksheet 20-3

Name _____ Station _____ Date _____

Battery Inspection

ASE Education Foundation Correlation

This lab worksheet addresses the following **MLR** tasks:

6.B.1. Perform battery state-of-charge test; determine necessary action. **(P-1)**

6.B.4. Inspect and clean battery; fill battery cells; check battery cables, connectors, clamps, and hold-downs. **(P-1)**

Description of Vehicle

Year _____ Make _____ Model _____

Engine _____ AT MT CVT PSD (circle which applies)

Battery group number _____   RC _____

CCA _____ CA _____ AH _____

Procedure

1. Battery open circuit voltage _____

   What is the battery state of charge based on the open circuit voltage? _____

   Based on the state of charge, what action(s) should be performed? _____

   _____

2. Inspect the case for damage and acid leaks. Note your findings. _____

   _____

3. Inspect and note the battery hold-down condition. _____

4. Inspect the terminals for fit, damage, and corrosion. Note your findings. _____

   _____

5. Inspect the cables for the following concerns:

   a. Damaged insulation   Yes _____   No _____

   b. Broken strands   Yes _____   No _____

   c. Secure connections   Yes _____   No _____

6. If the battery has removable cell caps, remove the caps and note the acid level in each cell.

   Cell 1_____   Cell 2 _____   Cell 3 _____   Cell 4 _____   Cell 5 _____   Cell 6 _____

   What may be indicated by low acid level in one or more cells? _____

   _____

   If necessary, add distilled water to each cell that is low. Do not overfill the cell.

                                  Instructor's check _____

7. Based on your inspection, what is the general condition of the battery? _____

   _____

   _____

8. Remove any tools and equipment and clean up the work area.

                                  Instructor's check _____

Lab Worksheet 20-4

Name _____ Station _____ Date _____

Maintain Memory

ASE Education Foundation Correlation

This lab worksheet addresses the following **MLR** tasks:

6.B.3. Maintain or restore electronic memory functions. **(P-1)**

6.B.8. Identify electrical/electronic modules, security systems, radios, and other accessories that require reinitialization or code entry after reconnecting vehicle battery. **(P-1)**

Description of Vehicle

Year _____ Make _____ Model _____

Engine _____ AT MT CVT PSD (circle which applies)

Procedure

1. List the systems on modern vehicles that use key-off battery power to maintain memory functions.

   _____

   _____

   _____

2. List any other systems or components that may be affected by disconnecting the battery or by the vehicle having a dead battery. _____

   _____

   _____

   _____

3. Describe two ways in which a vehicle's memory systems can be maintained with the battery disconnected.

   _____

   _____

4. Using a vehicle supplied by your instructor, check the following for memory presets and confirm settings.

_____ Clock   _____ Radio stations

_____ Seat/mirror positions   _____ Steering wheel

Instructor's check _____

5. Connect a memory-saving device to the vehicle following the manufacturer's procedures.

Instructor's check _____

6. If not already done, install fender covers. Disconnect the battery negative cable and leave it off for one minute. Reconnect the battery negative cable and remove the memory-saving device.

Instructor's check _____

7. Ensure the clock and radio station presets have been maintained.

Instructor's check _____

8. Remove your tools and equipment and clean up the work area.

Instructor's check _____

Lab Worksheet 20-5

Name _____ Station _____ Date _____

Charge Battery

ASE Education Foundation Correlation

This lab worksheet addresses the following **MLR** task:

6.B.5. Perform slow/fast battery charge according to manufacturer's recommendations. **(P-1)**

Description of Vehicle

Year _____ Make _____ Model _____

Engine _____ AT MT CVT PSD (circle which applies)

Procedure

1. List five safety precautions for charging a battery.

   _____

   _____

   _____

   _____

   _____

2. Locate and record the following battery information:

   CCA _____   A-h rating _____

   RC rating _____   Other rating _____

3. Perform a battery open circuit voltage test and record your results.

   Volts _____   State of charge _____

4. If using a constant voltage charger determine the correct charging rate and time based on the battery manufacturer's service information and the results of the open circuit voltage test. Connect the charger and set the charging rate and time.

Charging rate _____   Charging time _____

Instructor's check _____

5. If using a smart charger, connect the charger per the manufacturer's instructions and configure the charger.

Instructor's check _____

6. If using a constant voltage charger, periodically measure the battery voltage while charging. If voltage rises above 15.5 volts, turn the charger off to prevent battery overheating and damage. Explain how you will determine that the battery is fully charged. _____

_____

_____

7. Once the battery is charged, disconnect the charger and clean up the work area.

Instructor's check _____

Lab Worksheet 20-6

Name _____ Station _____ Date _____

Jump Start

ASE Education Foundation Correlation

This lab worksheet addresses the following **MLR** task:

6.B.6. Jump-start vehicle using jumper cables and a booster battery or an auxiliary power supply. **(P-1)**

Description of Vehicle

Year _____ Make _____ Model _____

Engine _____ AT MT CVT PSD (circle which applies)

Procedure

1. List five safety precautions for jump-starting a battery.

_____

_____

_____

_____

_____

2. Describe or draw the proper connection sequence to jump-start the dead battery.

Instructor's check _____

3. Install fender covers and connect the jumper cables in the correct sequence.

Instructor's check _____

4. Attempt to start the engine of the vehicle with the dead battery. Does the engine crank over?

Yes _____   No _____

5. If the engine does not crank, what are the possible causes? _____

_____

_____

6. To jump-start a vehicle with an auxiliary power supplier or booster pack, determine the correct way to connect the booster to the vehicle. Record the procedure. _____

_____

_____

Instructor's check _____

7. Connect the booster to the vehicle and attempt to start then engine. Does the engine crank over?

Yes _____   No _____

8. If the engine does not crank, what are the possible causes? _____

_____

_____

9. Disconnect the jumper-cables or booster and clean up the work area.

Instructor's check _____

Lab Worksheet 20-7

Name _____ Station _____ Date _____

HEV 12-Volt

ASE Education Foundation Correlation

This lab worksheet addresses the following **MLR** task:

6.B.9. Identify hybrid vehicle auxiliary (12v) battery service, repair, and test procedures. **(P-2)**

Description of Vehicle

Year _____ Make _____ Model _____

Engine _____ AT MT CVT PSD (circle which applies)

Procedure

1. Locate the 12V batter for this vehicle. Describe its location. _____
    _____

2. Inspect the battery decal for any specific service precautions. Note your findings. _____
    _____
    _____

3. Refer to the manufacturer's service information and note any special service precautions for the 12V battery
    for this vehicle. _____
    _____

4. List the procedures to test the 12V battery for this vehicle. _____
    _____
    _____
    _____

5. Explain why it is important to follow the correct procedures for inspecting and testing the 12V battery in this
    vehicle. _____
    _____
    _____

Lab Worksheet 20-8

Name _____ Station _____ Date _____

Starter Circuit Tests

ASE Education Foundation Correlation

This lab worksheet addresses the following **MLR** tasks:

6.C.1. Perform starter current draw test; determine necessary action. **(P-1)**

6.C.2. Perform starter circuit voltage drop tests; determine necessary action. **(P-1)**

Description of Vehicle

Year _____ Make _____ Model _____

Engine _____ AT MT CVT PSD (circle which applies)

Battery group number _____ CCA/CA _____

Procedure

1. Measure and record the open battery voltage. _____ volts

    Connect the starter testing equipment and record cranking volts and amps.

2. Cranking volts _____   Cranking amps _____

3. Compare your readings with the manufacturer's specifications.

    Cranking voltage specification _____   Cranking voltage specification _____

    Are the cranking volts and amps within specs?   Yes _____   No _____

4. Connect a DMM to the battery positive terminal and the starter positive cable connection. Crank the engine and record the voltage.

    _____ volts   Is this voltage drop acceptable?   Yes _____   No _____

5. Connect a DMM to the starter case and the battery negative terminal connection. Crank the engine and record the voltage.

   _____ volts   Is this voltage drop acceptable?   Yes _____   No _____

6. Connect a DMM to the starter control circuit terminal at the solenoid and the battery positive terminal connection. Crank the engine and record the voltage.

   _____ volts   Is this voltage drop acceptable?   Yes _____   No _____

7. Based on your testing, what is the condition of the starter and starter circuit? _____

   _____

8. What can cause higher than normal voltage drop readings for the starter motor? _____

   _____

   _____

9. Remove any tools and equipment and clean up the work area.

   Instructor's check _____

Lab Worksheet 20-9

Name _____ Station _____ Date _____

Starter Relays

ASE Education Foundation Correlation

This lab worksheet addresses the following **MLR** task:

6.C.3. Inspect and test starter relays and solenoids; determine necessary action. **(P-2)**

Description of Vehicle

Year _____ Make _____ Model _____

Engine _____ AT MT CVT (circle which applies)

Procedure

1. Describe the starting system circuits for this vehicle. _____

   _____

   _____

2. Describe how to test a relay. _____

   _____

   _____

3. If the vehicle is equipped with a starter relay, remove the relay and test it according to the manufacturer's service procedures. Note your results. _____

   _____

4. Describe how to test a starter solenoid. _____

   _____

   _____

4. Following the manufacturer's service information, test the starter solenoid and record your results.

   _____

   _____

5. Explain what concerns can be caused by a faulty starter relay. _____

   _____

   _____

6. Explain what concerns can be caused by a faulty starter solenoid. _____

   _____

   _____

7. Remove your tools and equipment and clean up the work area. _____

   Instructor's check _____

Lab Worksheet 20-10

Name _____ Station _____ Date _____

Test Relay

ASE Education Foundation Correlation

This lab worksheet addresses the following **MLR** task:

6.C.3. Inspect and test starter relays and solenoids; determine necessary action. **(P-2)**

Description of Vehicle

Year _____ Make _____ Model _____

Engine _____ AT MT CVT (circle which applies)

Relay manufacturer _____   Part number _____

Number of pins _____

Procedure

1. Describe the external condition of the relay. _____
_____

2. Draw the pin configuration and label the terminals as shown on the relay.

   If the relay terminals are not labeled, draw the pin configuration and label the terminals after testing.

3. Using a DMM, measure and record the relay coil resistance. _____

   Is the resistance within specs?   Yes _____   No _____   Unknown _____

4. Apply power and ground to the coil terminals. Does the relay click?

   Yes _____   No _____

   If no, what may this indicate? _____

5. With the relay coil energized, measure the resistance across the two remaining switch terminals. Resistance measurement _____

   What is indicated by this measurement? _____

6. With the relay coil energized, connect a power source and a test light to the switch terminals. Does the test light illuminate?   Yes _____   No _____

7. Based on your tests, what is the condition of the relay? _____
   _____

8. Remove any tools and equipment and clean up the work area.

   Instructor's check _____

Lab Worksheet 20-11

Name _____ Station _____ Date _____

Remove and Install Starter

ASE Education Foundation Correlation

This lab worksheet addresses the following **MLR** task:

6.C.4. Remove and install starter in a vehicle. **(P-1)**

Description of Vehicle

Year _____ Make _____ Model _____

Engine _____ AT MT CVT (circle which applies)

Procedure

1. Locate and record the manufacturer's procedures for removing and installing the starter on this vehicle.

   _____

   _____

2. Install fender covers. Disconnect the battery negative cable and ensure it cannot touch the battery once removed.

   Instructor's check _____

3. Remove any covers, shields, or other components that obstruct access to the starter. Remove the battery cable and control circuit connections at the starter.

   Instructor's check _____

4. Loosen and remove the starter mounting bolts. Remove the starter from the vehicle.

   Instructor's check _____

5. Locate the starter motor bench test or no-load test specifications.

   Bench test current draw specification _____

6. Following the manufacturer's service information, bench test the starter and record your results.

   Bench test current draw _____

7. Based on your test, what is the condition of the starter motor?

_____

8. Explain why many manufacturers specify the starter motor be bench tested. _____

_____

_____

9. Locate the starter mounting bolt and related fastener torque specs and record them.

   Torque specifications _____

10. Reinstall the starter and start the mounting bolts by hand. Ensure all bolts are threading properly before tightening. Use a torque wrench and torque the fasteners to specifications.

                         Instructor's check _____

11. Reconnect the battery and control circuit wiring. Do not overtighten the battery cable connection. Use a torque wrench and torque the fasteners to specifications.

                         Instructor's check _____

12. Reinstall any covers, shields, brackets or other components that were removed to access the starter.

                         Instructor's check _____

13. Reconnect the battery negative cable. Check starter operation by cranking the engine several times.

                         Instructor's check _____

14. Recheck your work and make sure all connections are tight and all components are in place and secure.

                         Instructor's check _____

15. Remove your tools and equipment and clean up the work area.

                         Instructor's check _____

Lab Worksheet 20-12

Name _____ Station _____ Date _____

Inspect Starter Circuit

ASE Education Foundation Correlation

This lab worksheet addresses the following **MLR** task:

6.C.5. Inspect and test switches, connectors, and wires of starter control circuits; determine necessary action. **(P-2)**

Description of Vehicle

Year _____ Make _____ Model _____

Engine _____ AT MT CVT (circle which applies)

Procedure

1. Refer to the manufacturer's service information and note any specific starter circuit testing procedures or specifications. _____

_____

_____

2. Many starter circuit concerns come from excessive voltage drop across switches and connections. Draw the starter circuit for this vehicle and note locations to test circuit voltage drops.

Instructor's check _____

3. With permission from your instructor, have an assistant crank the engine while you test the starter circuit.

Instructor's check _____

4. Begin by visually inspecting the wiring and components of the starter circuit. Note your findings.

5. If no problems are found, use a DMM to test voltage drop on the insulated power, insulated ground, and control circuits. Note your results of each circuit test.

Insulated power circuit voltage drops _____

Insulated ground circuit voltage drops _____

Control circuit voltage drops _____

6. Based on your testing, what is the condition of the starting circuit? _____

_____

_____

7. Remove your tools and equipment and clean up the work area.

Instructor's check _____

Lab Worksheet 20-13

Name _____ Station _____ Date _____

Idle-Stop

ASE Education Foundation Correlation

This lab worksheet addresses the following **MLR** task:

6.C.6. Demonstrate knowledge of an automatic idle-stop/start-stop system. **(P-3)**

Description of Vehicle

Year _____ Make _____ Model _____

Engine _____ AT MT CVT PSD (circle which applies)

Procedure

1. Locate and record any safety precautions for working on a vehicle with an idle-stop/stop-start system.

   _____

   _____

   _____

2. Describe the operation of the idle-stop and stop-start system used on this vehicle.

3. Does this vehicle have a method of disabling the idle-stop and stop-start system?

   Yes _____   No _____

   If yes, how is the system disabled? _____

   _____

4. Does the idle-stop and stop-start system automatically re-enable upon vehicle restart?

   Yes _____   No _____

5. Describe how the starting system and its components differ from those in a vehicle that does not have a idle-stop and stop-start system. _____

   _____

   _____

Lab Worksheet 20-14

Name _____ Station _____ Date _____

Test Charging Output

ASE Education Foundation Correlation

This lab worksheet addresses the following **MLR** task:

6.D.1. Perform charging system output test; determine necessary action. **(P-1)**

Description of Vehicle

Year _____ Make _____ Model _____

Engine _____ AT MT CVT PSD (circle which applies)

Procedure

1. Test battery voltage and record.   _____ volts

2. Inspect the drive belt for wear, damage, and tension. Record your findings. _____
   _____

3. Visually inspect the wiring to the generator. Record your findings. _____
   _____

4. Locate and record the generator output rating.   _____ amps

5. With a starting/charging system tester connected to the battery, record the charging output at idle with no loads turned on.
   _____ volts   _____ amps

6. With several loads turned on, record the idle output.
   _____ volts   _____ amps

7. Run the engine at 2,000 rpm and record the charging output with no loads.
   _____ volts   _____ amps

8. Load the generator until battery voltage reaches 12.6V and record the output.
   _____ amps

9. Based on your testing, what is the condition of the charging system? _____

_____

10. Remove any tools and equipment and clean up the work area.

Instructor's check _____

Lab Worksheet 20-15

Name _____ Station _____ Date _____

Generator Belt

ASE Education Foundation Correlation

This lab worksheet addresses the following **MLR** task:

6.D.2. Inspect, adjust, and/or replace generator (alternator) drive belts; check pulleys and tensioners for wear; check pulley and belt alignment. **(P-1)**

Description of Vehicle

Year _____ Make _____ Model _____

Engine _____ AT MT CVT (circle which applies)

Procedure

1. Locate and note the type of drive belt used.

   V-belt _____   Serpentine (multirib) belt _____

2. Describe how tension is applied to the generator belt. _____
   _____

3. Inspect the drive belt for wear and damage. Note your findings. _____
   _____
   _____

4. Locate the belt tension specifications and record them.   Specification _____

5. Using a belt tension gauge, measure and record the generator drive belt tension.

   Measured tension _____

6. How does the measured tension compare to the specification? _____

_____

7. Based on your inspection, what do you recommend? _____

_____

_____

8. Remove any tools and equipment and clean up the work area.

Instructor's check _____

Lab Worksheet 20-16

Name _____ Station _____ Date _____

Remove and Install Generator

ASE Education Foundation Correlation

This lab worksheet addresses the following **MLR** task:

6.D.3. Remove, inspect, and/or replace generator (alternator). **(P-2)**

Description of Vehicle

Year _____ Make _____ Model _____

Engine _____ AT MT CVT (circle which applies)

Procedure

1. Locate and record the manufacturer's procedures for removing and installing the generator on this vehicle.

   _____

2. Install fender covers. Disconnect the battery negative cable and ensure it cannot touch the battery once removed.

   Instructor's check _____

3. Remove any covers, shields, or other components that obstruct access to the generator. Loosen and remove the generator drive belt. Remove the battery cable and control circuit connections at the generator.

   Instructor's check _____

4. Loosen and remove the generator mounting bolts. Remove the generator from the vehicle.

   Instructor's check _____

5. Inspect the generator case, pulley, and connections. Note your findings. _____

   _____

   _____

6. If installing a new or remanufactured generator, make sure the new and old units match before attempting to install the new unit.

Instructor's check _____

7. Locate the generator mounting bolt and related fastener torque specifications and record them.

Torque specifications _____

8. Reinstall the generator and start the mounting bolts by hand. Ensure all bolts are threading properly before tightening. Use a torque wrench and torque the fasteners to specifications.

Instructor's check _____

9. Reinstall the drive belt onto the generator pulley. Make sure the belt fits fully down into the pulley grooves and the belt tension is correct.

Instructor's check _____

10. Reconnect the battery and control circuit wiring. Do not overtighten the battery cable connection. Use a torque wrench and torque the fasteners to specifications.

Instructor's check _____

11. Reinstall any covers, shields, brackets, or other components that were removed to access the generator.

Instructor's check _____

12. Reconnect the battery negative cable. Check generator operation by starting the engine and testing generator output.

Instructor's check _____

13. Recheck your work and make sure all connections are tight and all components are in place and secure.

Instructor's check _____

14. Remove your tools and equipment and clean up the work area.

Instructor's check _____

Lab Worksheet 20-17

Name _____ Station _____ Date _____

Voltage Drop Tests

ASE Education Foundation Correlation

This lab worksheet addresses the following **MLR** task:

6.D.4. Perform charging circuit voltage drop tests; determine necessary action. **(P-2)**

Description of Vehicle

Year _____ Make _____ Model _____

Engine _____ AT MT CVT PSD (circle which applies)

Procedure

Connect a starting/charging system tester to the vehicle.

1. Measure the voltage drop between the battery positive and the generator output terminal with the engine at idle.

    _____ Voltage drop no load

    _____ Voltage drop with load

2. Measure the voltage drop between the battery positive and the generator output terminal with the engine at 2,000 rpm.

    _____ Voltage drop

    _____ Voltage drop with load

3. Measure the voltage drop between the battery negative and the generator case with the engine at idle.

    _____ Voltage drop

    _____ Voltage drop with load

4. Measure the voltage drop between the battery negative and the generator with the engine at 2,000 rpm.

    _____ Voltage drop

    _____ Voltage drop with load

5. Did the voltage drop change with a load applied?   Yes _____   No _____

6. If yes, why do you think there was a change in the voltage drop amounts? _____
_____

7. Based on your testing, what is the condition of the charging system's wiring?
_____
_____

8. Remove any tools and equipment and clean up the work area.

Instructor's check _____

CHAPTER 21

Lighting and Electrical Accessories

Review Questions

1. Which of the following is not an advantage of LED lights?
   a. Increased bulb life
   b. Increased heat generation
   c. Decreased power consumption
   d. Decreased heat output

2. Door switches are used to either _____ a light circuit or to serve as an input for the _____.

3. Vehicles that have interior lights that slowly _____ and then go out use either the lighting or body _____ to control the lights.

4. Instrument panel illumination is controlled through the _____ switch.

5. When you are beginning the diagnosis of a customer's concern, begin by _____ the complaint.

6. *Technician A* says both incandescent and LED lights can be checked with an ohmmeter. *Technician B* says only incandescent lights can be checked with an ohmmeter. Who is correct?
   a. Technician A
   b. Technician B
   c. Both a and b
   d. Neither a nor b

7. A switch should show continuity when the switch is _____.

8. Explain why a more accurate test of a switch is performed by measuring its voltage drop during operation than by testing resistance only.

9. _____ means that there is a complete circuit or path between the meter's test leads.

10. In many vehicles, the door switches complete a _____ circuit through the switch itself.

11. Why is it important to use care and caution when removing interior lighting components?

12. Explain bulb trade numbers. _____

_____

_____

13. None of the IP bulbs operate, *Technician A* says the IP dimmer control may be faulty. *Technician B* says the IP dimmer may be turned down. Who is correct?

    a. Technician A

    b. Technician B

    c. Both a and b

    d. Neither a nor b

14. All of the following are types of exterior lights except:

    a. Back-up lights

    b. IP lights

    c. Turn signal lights

    d. Stoplights

15. Halogen insert bulbs have replaced _____ headlights because of their size and improved design flexibility.

16. HID lights use _____ AC to create an arc across two electrodes to produce light.

17. The _____ switch is used to switch between the low- and high-beam headlights.

18. DRL stands for _____.

19. Parking lamps are front, rear, or side marker bulbs operated by the _____ switch.

20. If the vehicle uses the same bulbs for the brake and _____, then the turn signal circuit connects to the brake light circuit.

21. Explain the operation of a bimetallic turn signal flasher unit. _____

_____

_____

22. Explain how using an incorrect wattage bulb can affect turn signal circuit operation.

_____

_____

_____

23. List four systems that often use the brake light switch as an input.

    a. _____

    b. _____

    c. _____

    d. _____

24. Define CHMSL. _____

    _____

25. The clear lights at the rear of the vehicle are used as:

    a. brake lights

    b. turn signal lights

    c. back-up lights

    d. all of the above

26. A vehicle with the headlight system shown in Figure 21-1 has no low-beam light operation. *Technician A* says a poor connection at terminal 2 of the dimmer switch could be the cause. *Technician B* says an open ground at G102 could be the cause. Who is correct?

    a. Technician A

    b. Technician B

    c. Both a and b

    d. Neither a nor b

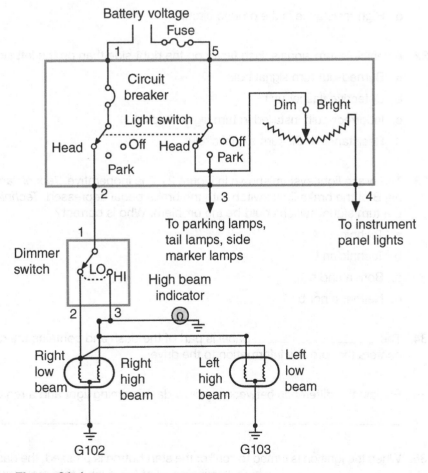

Figure 21-1

27. One headlight being brighter than the other can be caused by a _____ connection at the dim headlight.

28. Explain why it is important to not touch the glass bulb when replacing a halogen insert bulb. _____
_____

29. List three methods by which headlights are aimed. _____
_____
_____

30. Explain how to replace a typical parking light bulb. _____
_____

31. All of the rear parking lights are inoperative, but all other rear lights operate correctly. Which is the most likely cause?
    a. Defective brake light switch
    b. Open park light fuse
    c. All bulbs are burned out
    d. High resistance in the ground circuit

32. A vehicle's turn signals flash faster on the right side than on the left side. Which is the *least* likely cause?
    a. Burned-out turn signal bulb
    b. Defective flasher unit
    c. Incorrect bulb installed in turn signal socket
    d. Resistance in the turn signal circuit

33. The brake light system shown in Figure 21-2 is inoperative. *Technician A* says to check for power at terminal B of the brake light switch with the brake pedal depressed. *Technician B* says an open at terminal G of the turn signal switch could be the problem. Who is correct?
    a. Technician A
    b. Technician B
    c. Both a and b
    d. Neither a nor b

34. The _____ panel is part of the dash and contains the gauges, warning lights, and message centers that provide information to the driver.

35. Explain the difference between a yellow dash warning light and a red dash warning light. _____
_____

36. When the ignition is turned to "on" or the start button is pressed, the dash panel performs a _____
_____, which illuminates all of the dash lights momentarily.

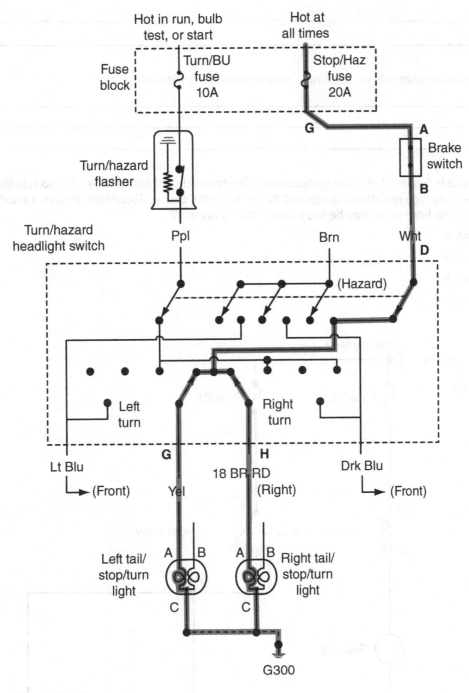

Figure 21-2

37. List nine common warning lights that warn a driver of a fault in a system. _____

_____

_____

_____

38. List three common methods of resetting a maintenance reminder light.

   a. _____

   b. _____

   c. _____

39. The horn circuit in Figure 21-3 is being discussed: The horns do not sound when the horn button is pressed. *Technician A* says high resistance at splice S105 may be the cause. *Technician B* says a short to ground at terminal 86 of the horn relay may be the cause. Who is correct?

   a. Technician A

   b. Technician B

   c. Both a and b

   d. Neither a nor b

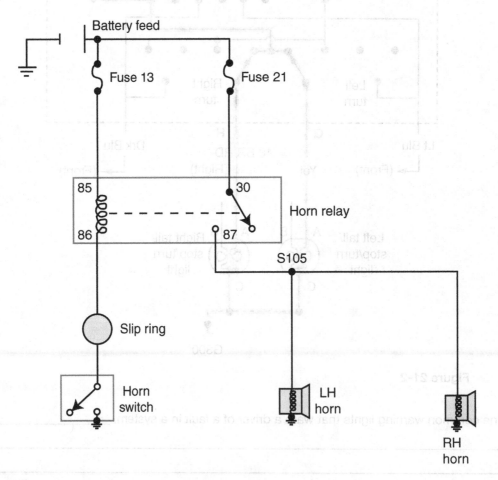

Figure 21-3

40. _____ systems use a remote to activate the power door locks, power sliding doors, and/or rear hatches.

41. *Technician A* says all remote keyless entry (RKE) systems use a key fob to unlock the doors. *Technician B* says some RKE systems use a smart key to unlock the doors and start the engine. Who is correct?
    a. Technician A
    b. Technician B
    c. Both a and b
    d. Neither a nor b

42. A vehicle has a complaint of intermittent power window and power door lock operation. *Technician A* says a faulty master switch may be the cause. *Technician B* says damaged wiring in the door jamb may be the cause. Who is correct?
    a. Technician A
    b. Technician B
    c. Both a and b
    d. Neither a nor b

43. The left-front window on a vehicle with BCM-controlled power windows does not operate from the door switch but does when controlled with a scan tool. Which is the most likely cause?
    a. Defective power window motor
    b. Defective power window switch
    c. Defective BCM
    d. Open in the power window motor wiring

Activities

1. The high beam lights do not operate in the circuit shown in Figure 21-4. Use the wiring diagram and the color codes shown to examine the circuit and its possible fault.

   Yellow for the component of concern

   Red for constant power to the component

   Orange for switched power to the component

   Green for a variable voltage to the component

   Black for the direct path to ground

   Blue for ground controlled by a switch or other control device

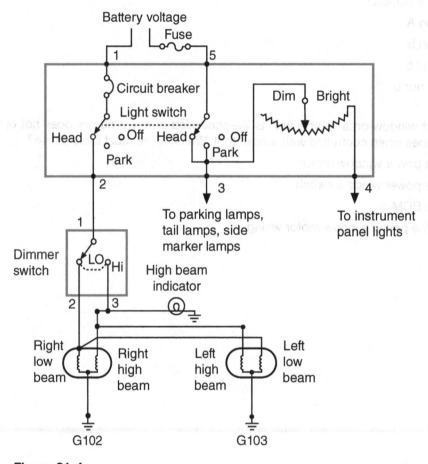

Figure 21-4

2. Obtain a selection of interior and exterior light bulbs from your instructor. Inspect each bulb for signs of damage, broken filaments, and loss of vacuum. Check the resistance of each bulb with a DMM and record your results.

a. Bulb type _____ Bulb number _____ Resistance _____

Condition _____

b. Bulb type _____ Bulb number _____ Resistance _____

Condition _____

c. Bulb type _____ Bulb number _____ Resistance _____

Condition _____

d. Bulb type _____ Bulb number _____ Resistance _____

Condition _____

e. Bulb type _____ Bulb number _____ Resistance _____

Condition _____

Lab Activity 21-1

Name _____ Date _____ Instructor _____

Year _____ Make _____ Model _____

Scan tool used _____

Check Module Status

1. Connect the scan tool and turn the ignition on. Does the scan tool power up correctly?

   Yes _____   No _____

   If no, what does this indicate? _____

2. Does the scan tool support a module or network status test?

   Yes _____   No _____

   If yes, what is the test called? _____

3. If the scan tool supports a module/network test, perform the test and record the results.

   _____

   _____

   _____

4. If the scan tool does not support module and network tests, attempt to communicate with modules known to be present on the vehicle. Record your results. _____

   _____

   _____

5. Based on your testing, are all modules active on the network?

   Yes _____   No _____

6. Do any of the modules have stored DTCs?   Yes _____   No _____

   If yes, record the DTCs. _____

7. Select an accessory system such as lighting for testing. System selected _____

8. List any bidirectional (active) command functions available in the system selected.

_____

_____

9. How can this tool capability be used to diagnose and repair a problem with the vehicle? _____

_____

Lab Worksheet 21-1

Name _____ Station _____ Date _____

Bulb ID

ASE Education Foundation Correlation

This lab worksheet addresses the following **MLR** task:

6.E.1. Inspect interior and exterior lamps and sockets including headlights and auxiliary lights (fog lights/driving lights); replace as needed. **(P-1)**

Description of Vehicle

Year _____ Make _____ Model _____

Engine _____ AT MT CVT PSD (circle which applies)

Procedure

1. Remove each type of bulb listed and record the bulb number. Bulb types are often found in the owner's manual or in the service information.

    a. Sealed beam head light _____

    b. Insert head light bulb—low beam _____

    c. Insert head light bulb—high beam _____

    d. Front turn signal/marker light _____

    e. Brake light _____

    f. License plate light _____

    g. Interior dome light _____

    h. Under dash or door panel light _____

2. Why do you think so many different types of bulbs are used in the vehicle? _____

    _____

    _____

3. What problems can result from replacing a bulb with an incorrect bulb type?

    _____

    _____

4. What problems can result from replacing an incandescent bulb with an LED bulb? _____

_____

_____

5. Remove any tools and equipment and clean up the work area.

Instructor's check _____

Lab Worksheet 21-2

Name _____ Station _____ Date _____

Inspect Bulbs with Scan Tool

ASE Education Foundation Correlation

This lab worksheet addresses the following **MLR** task:

6.E.1. Inspect interior and exterior lamps and sockets including headlights and auxiliary lights (fog lights/driving lights); replace as needed. **(P-1)**

Description of Vehicle

Year _____ Make _____ Model _____

Engine _____ AT MT CVT PSD (circle which applies)

Procedure

1. Scan tool used _____

2. Connect and configure the scan tool. Navigate to the system or module for lighting.

   System or module used _____

3. List the exterior lighting systems available _____
   _____

4. List the interior lighting circuits available _____
   _____

5. Select several lighting circuits to perform active commands. Turn each circuit on and note its operation.

   Circuit _____   Operation correct _____

   Circuit _____   Operation correct _____

   Circuit _____   Operation correct _____

6. If a bulb failed to operate, what would be your next step in diagnosis and why?

_____

_____

_____

7. If an entire lighting system failed to operate, what would be your next step in diagnosis?

_____

_____

8. Remove any tools and equipment and clean up the work area.

Instructor's check _____

Lab Worksheet 21-3

Name _____ Station _____ Date _____

Inspect Bulbs

ASE Education Foundation Correlation

This lab worksheet addresses the following **MLR** task:

6.E.1. Inspect interior and exterior lamps and sockets including headlights and auxiliary lights (fog lights/driving lights); replace as needed. **(P-1)**

Description of Vehicle

Year _____ Make _____ Model _____

Engine _____ AT MT CVT PSD (circle which applies)

Procedure

1. Turn on and check all exterior lights. Note the results of your inspection.

Headlights	OK _____	Not OK _____
Turn signals	OK _____	Not OK _____
Parking lights	OK _____	Not OK _____
Brake lights	OK _____	Not OK _____
Hazard lights	OK _____	Not OK _____
Other	OK _____	Not OK _____

2. Turn on and check all interior lights. Note the results of your inspection.

Under dash lights	OK _____	Not OK _____
Door lights	OK _____	Not OK _____
Dome lights	OK _____	Not OK _____
Instrument panel lights	OK _____	Not OK _____
Rear passenger area lights	OK _____	Not OK _____
Cargo area lights	OK _____	Not OK _____

3. Based on your inspection, what is the necessary action? _____

_____

_____

4. If any of the lights are inoperative, what is the next step to take? _____

_____

_____

5. Describe how to remove and check a nonfunctioning headlight. _____

_____

_____

6. Describe how to remove and check a nonfunctioning brake light. _____

_____

_____

7. Remove any tools and equipment and clean up the work area.

Instructor's check _____

Lab Worksheet 21-4

Name _____ Station _____ Date _____

Aim Headlights

ASE Education Foundation Correlation

This lab worksheet addresses the following **MLR** task:

6.E.2. Aim headlights. **(P-2)**

Description of Vehicle

Year _____ Make _____ Model _____

Engine _____ AT MT CVT PSD (circle which applies)

Procedure

1. Refer to the service information about how to properly check and aim the headlights on this vehicle. Summarize the information. _____

   _____

   _____

   _____

2. If using an alignment screen, position the vehicle as specified to the screen. Turn on the headlights and note the readings.

   Left side _____

   Right side _____

3. If the headlights need adjustment, turn the adjustment screws until the light shines at the specified readings.

   Instructor's check _____

4. If using headlight aiming equipment, set up the equipment per the manufacturer's instructions. Turn on the headlights and note the readings.

   Left side _____

   Right side _____

5. If the headlights need adjustment, turn the adjustment screws until the light shines at the specified readings.

Instructor's check _____

6. Explain why it is important for the headlights to be properly adjusted. _____

_____

_____

7. Remove your tools and equipment and clean up the work area.

Instructor's check _____

Lab Worksheet 21-5

Name _____ Station _____ Date _____

HID Safety

ASE Education Foundation Correlation

This lab worksheet addresses the following **MLR** task:

6.E.3. Identify system voltage and safety precautions associated with high-intensity discharge headlights. **(P-2)**

Description of Vehicle

Year _____ Make _____ Model _____

Engine _____ AT MT CVT PSD (circle which applies)

Procedure

1. Refer to the service information about HID service and repair safety precautions. Summarize the information. _____

_____

_____

_____

2. Explain why high-voltage is used on HID headlight systems. _____

_____

_____

3. Describe how to prepare the vehicle to service the HID headlight system. _____

_____

_____

4. What can result if the proper service procedures are not followed? _____

_____

_____

Instructor's check _____

Lab Worksheet 21-6

Name _____ Station _____ Date _____

Remove and Install Door Panel

ASE Education Foundation Correlation

This lab worksheet addresses the following **MLR** task:

6.E.5. Remove and reinstall door panel. **(P-1)**

Description of Vehicle

Year _____ Make _____ Model _____

Engine _____ AT MT CVT PSD (circle which applies)

Procedure

1. Refer to the service information for door panel removal and installation procedures. Summarize the information. _____

   _____

   _____

   _____

2. Locate and remove all fasteners holding the door panel in place. Note the types and locations for reinstallation.

   Instructor's check _____

3. Gently lift the door panel and separate it from the door. Disconnect any electrical connections between the panel and the door.

   Instructor's check _____

4. Once disconnected, store the door panel in a safe place to keep it clean and to prevent damage.

   Instructor's check _____

5. Inspect the door panel clips and all other hardware. Replace any damaged clips or fasteners.

   Instructor's check _____

6. To reinstall the door panel, reposition the panel close to the door and reconnect all electrical connections and other components you disconnected during removal.

Instructor's check _____

7. Carefully reinstall the door panel onto the door. Make sure all clips and fastener openings align before pressing the panel into place.

Instructor's check _____

8. Reinstall all fasteners and make sure the panel fits correctly against the door.

Instructor's check _____

9. Make sure all electrical switches operate correctly and that the door latch and locks operate.

Instructor's check _____

10. Remove your tools and equipment and clean up the work area.

Instructor's check _____

Lab Worksheet 21-7

Name _____ Station _____ Date _____

Describe RKE

ASE Education Foundation Correlation

This lab worksheet addresses the following **MLR** task:

6.E.6. Describe the operation of keyless entry/remote-start systems. **(P-3)**

Description of Vehicle

Year _____ Make _____ Model _____

Engine _____ AT MT CVT PSD (circle which applies)

Procedure

1. Describe the type of keyless entry system installed on this vehicle. _____
   _____
   _____

2. Use the service information to summarize how the system operates and what components are used.
   _____
   _____

3. Describe the operation of the remote-start system installed on this vehicle. _____
   _____
   _____

4. Use the service information to summarize the conditions required to enable the remote-start function.
   _____
   _____

5. Use the service information to summarize what will disable remote-start system operation.
   _____
   _____

Lab Worksheet 21-8

Name _____ Station _____ Date _____

Verify IP Operation

ASE Education Foundation Correlation

This lab worksheet addresses the following **MLR** task:

6.E.7. Verify operation of instrument panel gauges and warning/indicator lights; reset maintenance indicators. **(P-1)**

Description of Vehicle

Year _____ Make _____ Model _____

Engine _____ AT MT CVT PSD (circle which applies)

Procedure

1. Turn the ignition to ON or RUN and note the gauges and warning indicator lights on the instrument panel. Some indicator lights will remain on after 10 seconds while some will flash or turn off.

   Lights that remain on after 10 seconds: _____

   _____

   _____

   Lights that turn off within 10 seconds: _____

   _____

   _____

2. Note the operation of the gauges with the ignition on and the engine off. Do the gauges sweep or perform a self-test when the ignition is turned on?   Yes _____   No _____

3. What do the gauges display after 10 seconds with the key on? _____

   _____

   _____

4. With permission from your instructor, start the engine.

   Instructor's check _____

5. Note any lights that remain of once the engine is running. _____

_____

_____

6. What is indicated by these lights remaining on? _____

_____

_____

7. With the engine running, note the readings displayed by the gauges. _____

_____

_____

8. Based on your inspection, describe any concerns present related to the indicator lights and gauges.

_____

_____

Instructor's check _____

9. Turn the ignition off and connect a scan tool to the DLC. Turn the ignition on and configure the scan tool to communicate with the vehicle's body or instrument panel module.

Instructor's check _____

10. Does the vehicle provide a test of the instrument panel lights and gauges?

Yes _____   No _____

11. If Yes, perform the test. Note that all lights illuminate and that all gauges operate.

Instructor's check _____

12. If no testing is available, check for any stored DTCs related to the instrument cluster or its inputs. Note your findings. _____

_____

_____

13. Based on your testing, what actions are necessary? _____

_____

_____

14. Turn off the scan tool, turn off the ignition, and disconnect the scan tool. Remove your tools and equipment and clean up the work area.

Instructor's check _____

Lab Worksheet 21-9

Name _____ Station _____ Date _____

High-Voltage HID, Ignition, Injection

ASE Education Foundation Correlation

This lab worksheet addresses the following **RST** task:

1.14. Demonstrate awareness of the safety aspects of high voltage circuits (such as high intensity discharge (HID) lamps, ignition systems, injection systems, etc.).

Procedure

1. Describe what dangers are present from working on or near HID lighting systems. _____
   _____
   _____

2. Summarize how to make a HID headlight system safe before working on the system. _____
   _____
   _____

3. Explain the hazards present when working near or on secondary ignition system components. _____
   _____
   _____

4. List the ignition and fuel system components you need to use caution when working around when the engine is running. _____
   _____

Instructor's check _____

CHAPTER 22

Engine Performance Principles

Review Questions

1. The dominant types of engines in use today are the gasoline-powered _____ cycle engine and the engine.

2. In a gasoline-powered engine, the air and fuel mixture is ignited by a high-voltage spark delivered to the _____ chamber.

3. A diesel engine _____ air in the cylinder so much that the heat ignites the fuel injected into the cylinder.

4. At their most basic, internal combustion engines are _____ pumps.

5. At sea level, _____ pressure is 14.7 pounds per square inch.

6. Atmospheric pressure _____ as altitude increases.

7. All of the following statements about pressure and vacuum are correct except:

   a. downward piston movement increases pressure in the cylinder.
   b. the lower pressure in the engine is called vacuum.
   c. the difference between outside and inside the cylinder causes air to flow into the engine.
   d. all of the above.

8. The pressure inside the engine that is lower than atmospheric pressure is called _____.

9. A tire pressure gauge is calibrated to read _____ pressure at normal atmospheric pressure.

10. Explain why atmospheric pressure and vacuum are necessary for an engine to run.

_____

_____

_____

_____

11. Both gasoline and diesel fuel are made of _____.

12. Gasoline contains many _____ to make it usable as a fuel.

13. Diesel fuel can contain certain types of _____, which not only survive in the fuel but also actually feed on it.

14. To be able to extract the chemical energy stored in gasoline, the air–fuel mixture is _____, forcing the molecules closer together.

15. How much the air–fuel charge is compressed depends on the _____ ratio of the engine.

16. Uncontrolled _____ causes reduced power, poor performance, increased emissions, and in some cases, engine damage.

17. The reciprocating motion of the pistons must be converted into _____ motion.

18. List the four cycles of the internal combustion engine in order.
    a. _____
    b. _____
    c. _____
    d. _____

19. All current production gasoline engines for automotive use in the United States are _____ -cycle, _____-cooled, _____ ignition engines.

20. All of the following statements about diesel engines are correct except:
    a. automotive diesels are liquid-cooled.
    b. diesel engines do not use spark plugs.
    c. diesel engines are smaller and lighter than gasoline engines.
    d. diesel engines use four-cycle operation.

21. In an Atkinson cycle engine, the _____ stroke is extended, allowing some of the air–fuel mixture to move back up into the intake manifold.

22. The type of internal combustion engine that does not use reciprocating parts is called the
    a. diesel engine
    b. rotary engine
    c. Miller engine
    d. All engines use reciprocating parts

23. All of these construction materials are used in modern engine design to save weight except:

   a. aluminum

   b. plastic

   c. cast iron

   d. magnesium

24. Identify the components labeled in Figure 22-1.

   A. _____   B. _____

   C. _____   D. _____

   E. _____   F. _____

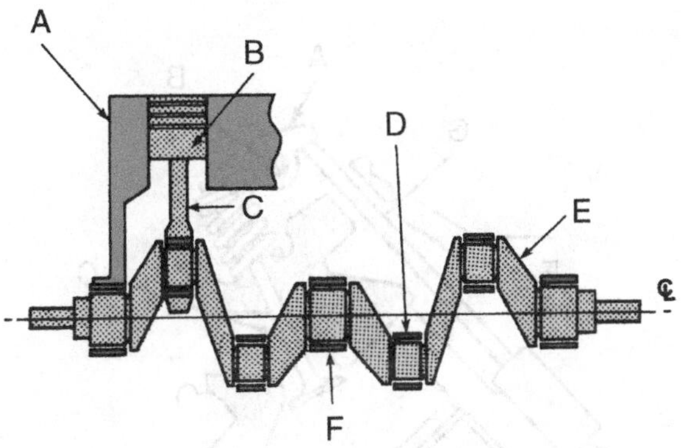

Figure 22-1

25. List six components of the engine bottom end.

   a. _____

   b. _____

   c. _____

   d. _____

   e. _____

   f. _____

26. Which of the following is not a function of the piston?

   a. Compress the air–fuel mixture

   b. Transfer movement to the crankshaft

   c. Seal the top of the combustion chamber

   d. Draw air into the cylinder

27. In an overhead valve engine, the camshaft is located in the _____.

28. List six components of the engine top end. _____

_____

_____

_____

_____

29. Label the parts shown in Figure 22-2.

A. _____   B. _____

C. _____   D. _____

E. _____   F. _____

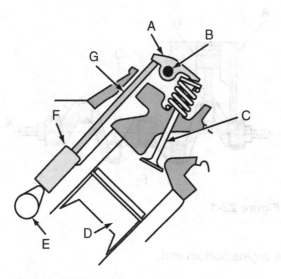

Figure 22-2

30. Explain the operation of the overhead valvetrain. _____

_____

_____

_____

_____

31. Explain the operation of the overhead camshaft valvetrain. _____

_____

_____

_____

_____

32. Getting the air into the engine is the responsibility of the _____ system.

33. During low-speed operation, _____ intake runners improve power by speeding up the airflow.

34. A turbocharger is driven by the:

   a. drive belt

   b. electric motor

   c. exhaust gases

   d. intake airflow

35. Turbocharged and supercharged engines use a(n) _____ to remove some of the heat from the air entering the engine.

36. List four functions of the exhaust system. _____

_____

_____

_____

37. The lubrication system is used to remove _____, clean the inside of the engine, and trap _____.

38. On some diesel engines, engine oil may be used to open and close the _____.

39. Explain three functions of the cooling system. _____

_____

_____

_____

40. Coolant is a mixture of _____-_____ and water.

41. Explain why a pressurized cooling system is used on modern engines.

_____

_____

_____

_____

42. Label the parts of the cooling system shown in Figure 22-3.

A. _____   B. _____

C. _____   D. _____

E. _____   F. _____

G. _____

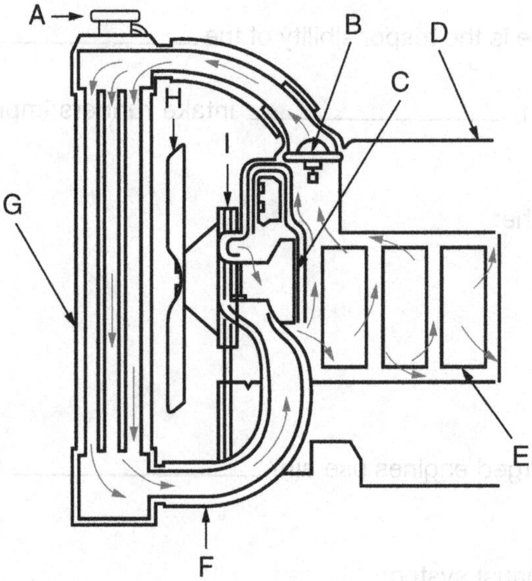

Figure 22-3

43. The _____ system on a gasoline-powered engine provides the heat to ignite the air–fuel mixture in the combustion chamber.

44. All modern gasoline engines use _____ to deliver the gasoline to the cylinders.

45. Describe in detail what happens during the intake stroke of a gasoline engine.

_____

_____

_____

_____

46. Describe in detail what happens during the compression stroke of a gasoline engine.

_____

_____

_____

_____

47. Describe in detail what happens during the combustion stroke of a gasoline engine.

_____

_____

48. Describe in detail what happens during the exhaust stroke of a gasoline engine.

_____

_____

_____

49. Horsepower is the rate of the amount of _____ performed in a specific amount of time.

50. Gasoline-powered engines are approximately _____ efficient.
    a. 80 percent
    b. 35 percent
    c. 20 percent
    d. 10 percent

51. Explain the fuel efficiency advantage of the Atkinson cycle engine.

_____

_____

52. Hybrid-electric vehicles (HEVs) combine ICEs and powerful electric motor and generators to _____ the vehicle, recapture _____ energy, increase fuel economy, and reduce _____ emissions.

Activities

I. Displacement

Engine size is determined by cylinder displacement, which is the total volume of the cylinders and combustion chambers. To understand displacement, examine the volume of a basic cylinder, as shown in Figure 22-4. When the piston is at BDC, the cylinder can hold a certain amount of liquid, represented by V. To determine the value of V, we need to know the radius of the cylinder and its depth. Radius is one-half of the diameter, which is the distance across the cylinder from side-to-side. Depth is the total height from the top edge to the bottom of the cylinder, which is also the top of the piston.

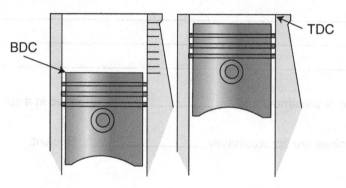

Figure 22-4

If the cylinder is 3.5 in. (88.9 mm) in diameter and 3.5 in. (88.9 mm) deep, this would equal a bore of 3.5 in. and a stroke of 3.5 in. Bore and stroke are the terms used to describe cylinder diameter and depth, as shown in Figure 22-5. To calculate the volume, the bore of 3.5 in. is converted to radius. Since radius is one-half of diameter (or bore), then a bore of 3.5 in. has a radius of 1.75 in. (44.45 mm). Using the formula $V = \pi r^2 h$, where V is the total volume, π is the value of pi, or 3.14159, r^2 is the cylinder radius squared, and h is the height (or stroke) of the cylinder, the volume can be determined. Our equation will look like $V = 3.14159 \times 1.75^2 \times 3.5$. The cylinder volume, or displacement, equals about 33.67 cubic inches, or 33.67 In.^{3} In metric measurements, the cylinder displaces about 552 cubic centimeters or 552 cc. If our engine has eight cylinders, then we multiply 8×33.67 In.^{3} to get a total engine displacement of 269.39 cubic inches or 4416 cc.

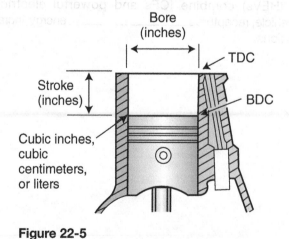

Figure 22-5

Another method of calculating displacement is by using the formula: 0.785 Bore × Stroke × Number of cylinders. Since 0.785 is one-quarter of π, some people find it easier to remember this way since the actual bore measurement can be used.

To measure displacement, obtain from your instructor a cylindrical container that you can measure for diameter and depth. Using a telescoping gauge and depth gauge, measure the inside diameter and depth of the container and record. Calculate the volume of the container using the formulas above.

1. Inside diameter _____ Depth _____ Volume _____

Using the volume you found, multiply the total by 4, 6, and 8 to get an idea of what the amount equates to for a four-cylinder, six-cylinder, and eight-cylinder engine.

2. Four-cylinder _____ Six-cylinder _____ Eight-cylinder _____

Compare these numbers with the engine sizes found in modern cars and light trucks.

3. How does the displacement compare to the external size of an engine? _____

4. What are some factors that limit engine displacement? _____

5. Can engine wear affect displacement? Why/how? _____

6. If an engine is bored oversize to correct for cylinder wear, what effect will this have on displacement?

_____

7. How would carbon buildup on a piston affect displacement? _____

II. Compression Ratio

Compression ratio refers to the volume of the cylinder when the piston is at BDC compared to the volume when the piston is at TDC, as shown in Figure 22-6. The difference in volume is the compression ratio.

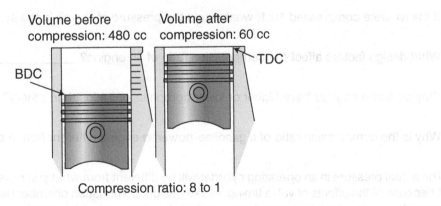

Volume before compression: 480 cc   Volume after compression: 60 cc

BDC   TDC

Compression ratio: 8 to 1

Figure 22-6

Obtain from your instructor a syringe that holds 100 cc of liquid and has graduated markings along the length marking every 10 cc, like that shown in Figure 22-7. Pull the plunger back, filling the syringe with 100 cc of air. This is similar to filling the cylinder of an engine when the piston has moved from TDC to BDC on the intake stroke.

Now, cap the opening of the syringe and push the plunger back in, compressing the trapped air inside. Notice that as you compress the air, it becomes more difficult to push the plunger. If you can compress the original 100 cc volume of air until the plunger reaches the 10 cc mark, you have compressed the air by a factor of 10, or by a ratio of 10:1. By taking the original volume of 100 cc and reducing it down to 10 cc, you have compressed the air 10 times. Compression ratio = volume 1/volume 2 or 100/10 = 10 or 10:1.

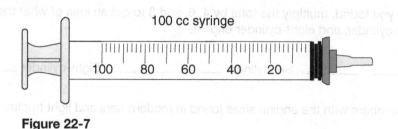

100 cc syringe

Figure 22-7

1. Why did the plunger become harder to push as the volume of air decreased? _____

2. What would the compression ratio in the syringe be if you compressed the air from 100 cc to 50 cc?

_____

3. If you compressed the air in the syringe to the 20 cc mark, what would the compression ratio be?

_____

For a simple experiment, Boyle's Law can be used to determine the pressure in the syringe. P = pressure, P_i = initial pressure, or 14.7 psi, or atmospheric pressure at sea level, V_i is the initial volume, in this case 100 cc, and V_f is the final volume. Use the formula for final pressure or $P_f = P_i V_i / V_f$ and a compression ratio of 2:1 (100 cc to 50 cc):

4. What would be air pressure in the syringe? _____

5. If the air were compressed 10:1, what would the pressure of the air in the syringe be? _____

6. What design factors affect the compression ratio of an engine? _____

7. Why do some engines have higher or lower compression ratios than others? _____

8. Why is the compression ratio of a gasoline-powered engine different from a diesel engine? _____

The actual pressure in an operating cylinder will be different from what you have experienced with this experiment because of the effects of valve timing, intake design, combustion chamber design, and other factors. These experiments do not take into account combustion chamber size, which also has an effect on displacement and compression.

Lab Activity 22-1

Name _____ Date _____ Instructor _____

Year _____ Make _____ Model _____

Determine Engine Design and Construction

Determine the following:

1. Engine size _____ cubic inches _____ liters

2. Engine type _____ V-block _____ Inline _____ Boxer

   _____ W-block _____ Rotary _____ Other

3. Cylinder bore _____ Stroke _____

4. Compression ratio _____ Firing order _____

5. Horsepower _____ Torque _____ at _____ rpm

6. Valvetrain type _____ OHV _____ OHC _____ DOHC

7. Fuel type _____ Gasoline _____ Diesel _____ Other

8. Fuel delivery system _____ Port fuel injection _____ Direct injection

   _____ Central injection _____ Other

9. Block construction material _____

10. Cylinder head construction material _____

11. Intake manifold construction material _____

Lab Worksheet 22-1

Name _____ Station _____ Date _____

Research Vehicle Information

ASE Education Foundation Correlation

This lab worksheet addresses the following **MLR** tasks:

1.A.1 - 8.A.1 Research vehicle service information, including fluid type, vehicle service history, service precautions, and technical service bulletins. **(P-1)**

Description of Vehicle

Year _____ Make _____ Model _____

Engine _____ AT MT CVT PSD (circle which applies)

Procedure

1. Describe why the vehicle was brought in to the shop. _____

    _____

2. Locate and describe any service history for this vehicle. _____

    _____

    Is any of the service history related to why the vehicle was brought in?   Yes _____   No _____

3. Using the service information, locate and describe any relevant service precautions. _____

    _____

4. Using the service information, locate and describe any relevant technical service bulletins (TSBs).

    _____

    _____

CHAPTER 23

Engine Mechanical Testing and Service

Review Questions

1. Modern engine control systems cannot correct for _____ compression or incorrect timing problems.

2. Before any attempt is made to correct a performance problem, the engine's _____ condition must first be verified.

3. When diagnosing a fluid leak, a _____ light may be used with a special dye to make locating the leak easier.

4. Some vacuum gauges can measure both vacuum and _____ pressure.

5. Scopes display _____ and _____ over time, making them very useful for testing sensors.

6. Explain why you should let the engine cool down before performing services.

   _____

   _____

7. List five different types of fluid leaks that are possible from around the engine compartment.

   a. _____

   b. _____

   c. _____

   d. _____

   e. _____

8. Describe how to distinguish between the different types of fluids you listed in question 7.

   _____

   _____

   _____

   _____

9. Explain how to perform an engine oil pressure test. _____

_____

_____

_____

10. *Technician A* says a vacuum leak may affect all cylinders equally. *Technician B* says a vacuum leak may only affect one cylinder. Who is correct?

    a. Technician A

    b. Technician B

    c. Both A and B

    d. Neither A nor B

11. A vacuum leak will make the air/fuel ratio:

    a. Rich

    b. Lean

    c. 14.7:1

    d. There will be no effect on the air/fuel ratio

12. A power balance test is performed to:

    a. Determine why a cylinder has low compression

    b. Determine which cylinder is not producing its share of power

    c. Locate a vacuum leak

    d. None of the above

13. On some vehicles, a _____ tool is used to perform a power balance test.

14. All of the following can cause a compression leak except:

    a. Leaking head gasket

    b. Burnt exhaust valve

    c. Broken oil control ring

    d. Cracked cylinder head

15. An engine has a low compression reading on one cylinder. *Technician A* says a wet test should be performed to determine if the problem is with the valves or the rings. *Technician B* says a wet test will determine if the head gasket is leaking. Who is correct?

    a. Technician A

    b. Technician B

    c. Both A and B

    d. Neither A nor B

16. A _____ compression test is performed using a current clamp and a scope.

17. The waveform shown in Figure 23-1 was captured during a relative compression test on an eight-cylinder engine. *Technician A* says this indicates normal cranking compression readings. *Technician B* says a problem is indicated in one cylinder. Who is correct?

    a. Technician A

    b. Technician B

    c. Both A and B

    d. Neither A nor B

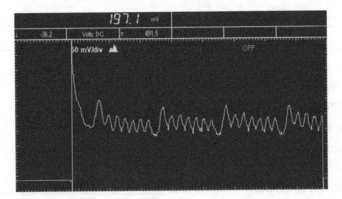

Figure 23-1

18. Explain what conditions can be diagnosed using a running compression test.

    _____

    _____

    _____

    _____

19. A running compression test is often used to pinpoint a problem with:

    a. Worn cam lobes

    b. Worn piston rings

    c. Bent valves

    d. Restricted exhaust

20. Which of the following types of problems is *not* diagnosed with a cylinder leakage test?

    a. Worn rings

    b. Worn cam lobes

    c. Leaking head gasket

    d. Bent valve

21. Which of the following tests is performed to determine if a cylinder is not producing power?

    a. Compression test

    b. Vacuum test

    c. Power balance test

    d. Cylinder leakage test

22. Bubbles are present in the cooling system during a cylinder leakage test. Describe what can cause this to occur.

    _____

    _____

23. Match the color of exhaust smoke to its cause for a gasoline engine.

    Blue   Combustion/cooling system leak

    White   Excessive fuel

    Black   Excessive oil in cylinder

24. Describe what a typical valvetrain noise sounds like. _____

    _____

    _____

25. An engine has an unusual knocking noise during operation. Which of the following tests should be performed?

    a. Vacuum test

    b. Oil pressure test

    c. Compression test

    d. Power balance test

26. Induction and vacuum leaks usually cause a high-pitched _____ sound.

27. Explain two problems associated with broken powertrain mounts. _____

    _____

    _____

    _____

28. List four possible causes of engine vibration. _____

    _____

    _____

    _____

29. List three common exhaust system problems. _____
_____
_____
_____

30. Explain how to test for a restricted exhaust system. _____
_____
_____
_____
_____

31. Describe five factors involved in deciding whether to rebuild or replace an engine.
    a. _____
    b. _____
    c. _____
    d. _____
    e. _____

32. Describe the differences between minor and major engine concerns. _____
_____
_____
_____

33. List three reasons why a shop may not perform major engine repairs that require rebuilding an engine.

34. Some engines require periodic valve _____ adjustments.

35. Define valve lash and why it is important that it be checked and adjusted.
_____
_____
_____
_____

36. To measure and set valve lash, a _____ gauge is used between the rocker arm and the valve.

37. *Technician A* says excessive valve lash may result in a tapping noise with the engine running. *Technician B* says if the valve lash is too small, the valve may stay open when the engine is cold. Who is correct?

    a. Technician A

    b. Technician B

    c. Both A and B

    d. Neither A nor B

38. Describe what is meant by the term *interference engine*. _____

    _____

    _____

39. _____ are used between a moving and a nonmoving part, and are used between two nonmoving parts.

40. *Technician A* says steel gasket scrapers are acceptable to use on all types of engine components. *Technician B* says gasket cleaning discs can safely be used on all types of gasket surfaces. Who is correct?

    a. Technician A

    b. Technician B

    c. Both A and B

    d. Neither A nor B

Lab Worksheet 23-1

Name _____ Station _____ Date _____

Leak Inspection

ASE Education Foundation Correlation

This lab worksheet addresses the following **MLR** task:

1.A.3. Inspect engine assembly for fuel, oil, coolant, and other leaks; determine necessary action. **(P-1)**

Description of Vehicle

Year _____ Make _____ Model _____

Engine _____ AT MT CVT PSD (circle which applies)

Procedure

1. Different automotive fluids are different colors to help with identification but the colors can change over time. Below are examples of fluid colors when new. Over time, engine oil, ATF, P/S, brake fluid, and coolants become darker and can look similar to each other.

   - Red—automatic transmission fluid (also used in some older power steering units)

   - Green—coolant

   - Orange—coolant

   - Blue—coolant or windshield washer fluid

   - Yellow—coolant or windshield washer fluid

   - Clear—condensation from the air-conditioning system (normal)

   - Dark brown or black—engine oil

   - Amber or light brown—brake fluid

2. Raise the hood, install a fender cover, and inspect these areas where oil is likely to leak:

Valve covers	OK _____	Not OK _____
Intake manifold area	OK _____	Not OK _____
Oil pressure-sending unit	OK _____	Not OK _____
Cylinder head gaskets	OK _____	Not OK _____

3. Safely lift and support the vehicle and carefully inspect the underside of the engine. Note your findings.

_____

_____

4. Over time, leaks can cover the engine and other parts of the vehicle. You may need to clean the area or add a UV dye to help locate the leak. Before using a UV light kit, read and summarize the safety warnings included with the UV light kit. _____

_____

5. If using UV dye, add the dye to the engine oil and run the engine for 15 minutes. Shut the engine off and using a black light and UV safety glasses, look for signs of the leak by looking for the yellow/green areas

   highlighted by the dye. Describe your findings. _____

_____

_____

6. Based on your inspection, what is the necessary action? _____

_____

7. Remove any tools and equipment and clean up the work area.

_____

Instructor's check _____

Lab Worksheet 23-2

Name _____ Station _____ Date _____

Identify Cylinder Head and Valve Train

ASE Education Foundation Correlation

This lab worksheet addresses the following **MLR** task:

1.B.2. Identify components of the cylinder head and valve train. **(P-1)**

Description of Vehicle

Year _____ Make _____ Model _____

Engine _____ AT MT CVT PSD (circle which applies)

Procedure

1. Using the service information, locate an exploded view or part explosion of the cylinder head.

    Instructor's check _____

2. Which type of valve train is this?   OHV   OHC   DOHC

3. Locate the following parts of the cylinder head and valve train. Note, not all cylinder heads are the same and may not contain all of these components.

    a) Intake port   _____   j) Hydraulic lifters   _____

    b) Exhaust port   _____   k) Valve seats   _____

    c) Combustion chamber   _____   l) Valve guides   _____

    d) Intake valve(s)   _____   m) Rocker arms   _____

    e) Exhaust valve(s)   _____   n) Push rods   _____

    f) Rocker arm boss or stud _____   o) Valve springs   _____

    g) Cam bearing journals   _____   p) Valve spring retainers _____

    h) Rocker arm shaft   _____   q) VVT cam phasor   _____

    i) Coolant passage   _____   r) Cam lobe   _____

Lab Worksheet 23-2

Name _____ Station _____ Date _____

Identify Cylinder Head and Valve Train

ASE Education Foundation Correlation

This lab worksheet addresses the following MLR task:

1.D.2. Identify components of the cylinder head and valve train. (P-1)

Description of Vehicle

Year _____ Make _____ Model _____

Engine _____ AT MT CVT PSD (circle when applies)

Procedure

1. Using the service information, locate an exploded view or part explosion of the cylinder head.
Instructor's check _____

2. Which type of valve train is this?   OHV   OHC   DOHC

3. Locate the following parts of the cylinder head and valve train. Note, not all cylinder heads are the same and may not contain all of these components.

a) Intake port _____

b) Exhaust port _____

c) Combustion chamber _____

d) Intake valve(s) _____

e) Exhaust valve(s) _____

f) Rocker arm boss or stud _____

g) Cam bearing journals _____

h) Rocker arm shaft _____

i) Coolant passage _____

j) Hydraulic lifters _____

k) Valve seats _____

l) Valve guides _____

m) Rocker arms _____

n) Push rods _____

o) Valve springs _____

p) Valve spring retainers _____

q) VVT cam ; chain _____

r) Cam lobes _____

Lab Worksheet 23-3

Name _____ Station _____ Date _____

Vacuum Tests

ASE Education Foundation Correlation

This lab worksheet addresses the following **MLR** task:

8.A.2. Perform engine absolute manifold pressure tests (vacuum/boost); document results. **(P-2)**

Description of Vehicle

Year _____ Make _____ Model _____

Engine _____ AT MT CVT PSD (circle which applies)

Procedure

1. Connect a vacuum gauge to a suitable vacuum port on the engine. Describe the location of the port.

   _____

2. With the ignition and/or fuel system disabled, crank the engine and record the vacuum reading.
   Cranking vacuum reading _____
   Reconnect the ignition and/or fuel system.

3. Start the engine and allow it to idle. What is the idle vacuum reading? _____
   a. Is the gauge showing a steady needle? Yes _____   No _____
   b. What would a bouncing or fluctuating needle indicate? _____

   _____

4. Quickly snap the throttle wide open and then closed. Note the readings at WOT and on deceleration
   WOT _____   Decel _____

5. Increase the engine speed to 2,000rpm and hold. Note the vacuum reading.

6. Does the vacuum reading decrease after one minute?   Yes _____   No _____

7. If the vacuum reading dropped at 2,000rpm, what would that indicate? _____

_____

_____

8. Based on your testing, what is the condition of the engine? _____

_____

_____

9. Remove any tools and equipment and clean up the work area.

Instructor's check _____

Lab Worksheet 23-4

Name _____ Station _____ Date _____

Power Balance Test

ASE Education Foundation Correlation

This lab worksheet addresses the following **MLR** task:

8.A.3. Perform cylinder power balance test; document results. **(P-2)**

Description of Vehicle

Year _____ Make _____ Model _____

Engine _____ AT MT CVT PSD (circle which applies)

Procedure

1. Obtain a scan tool applicable for the vehicle being tested.

   Scan tool used _____

2. Connect the scan tool to the DLC and enter into powertrain diagnostics. Locate the power balance test.

3. Initial engine rpm _____

4. Follow the procedure indicated by the scan tool, disable each cylinder, and note the rpm drop:

   Cylinder 1 _____   Cylinder 5 _____

   Cylinder 2 _____   Cylinder 6 _____

   Cylinder 3 _____   Cylinder 7 _____

   Cylinder 4 _____   Cylinder 8 _____

5. What is the rpm difference between the weakest and strongest cylinders? _____

6. Based on the test results, what is the necessary action? _____

   _____

   _____

7. Remove any tools and equipment and clean up the work area.

   Instructor's check _____

Lab Worksheet 23-5

Name _____ Station _____ Date _____

Cranking Compression Test

ASE Education Foundation Correlation

This lab worksheet addresses the following **MLR** task:

8.A.4. Perform cylinder cranking and running compression tests; document results. **(P-2)**

Description of Vehicle

Year _____ Make _____ Model _____

Engine _____ AT MT CVT PSD (circle which applies)

Procedure

1. Disable the fuel and ignition systems. Explain how this is accomplished. _____
    _____

    Warning! Hybrid vehicles require special procedures for conducting compression tests. Do not try to perform this test on a hybrid without following the manufacturer's precautions and procedures.

2. Use an air blow gun to clean around the spark plugs. Remove all the spark plugs.

3. Block the throttle open. Explain why the throttle should be wide open when performing a compression test. Note: Refer to the service information for procedures if vehicle is equipped with electronic throttle control.
    _____

    _____

4. Install a battery charger and set to a low charge rate. Explain why a battery charger should be used during a compression test. _____
    _____

5. Install the compression gauge into cylinder number 1. Crank the engine for five seconds and note the compression gauge reading. Record the reading in the space below.

Dry test	Wet test	Dry test	Wet test
Cylinder number	_____	Cylinder number	_____
Cylinder number	_____	Cylinder number	_____
Cylinder number	_____	Cylinder number	_____
Cylinder number	_____	Cylinder number	

6. If a cylinder's compression reading is more than 10 percent lower than the others, a wet test should be performed. Squirt a small amount of oil into the low cylinder and repeat the compression test.

7. Based on your testing results, what is the condition of the engine? _____

_____

_____

8. Remove any tools and equipment and clean up the work area.

Instructor's check _____

Lab Worksheet 23-6

Name _____ Station _____ Date _____

Running Compression Test

ASE Education Foundation Correlation

This lab worksheet addresses the following **MLR** task:

8.A.4. Perform cylinder cranking and running compression tests; document results. **(P-2)**

Description of Vehicle

Year _____ Make _____ Model _____

Engine _____ AT MT CVT PSD (circle which applies)

Procedure

1. Record the cylinder firing order: _____

2. Remove the spark plug from the cylinder being tested. Install a compression gauge in the spark plug hole. To monitor actual cylinder pressures, remove the Schrader valve from the compression tester. This allows the needle to rapidly bounce with cylinder pressure changes.

    Note: Damage to the gauge can result if the needle is pulled against the stop pin during this test. Depending on the type of gauge being used, leave the Schrader valve in the compression tester and use the bleed valve to bleed air from the gauge during the test.

3. Start the engine and allow the reading to stabilize. Needle bounce is normal for a running compression test. Cylinder running pressure _____

4. Snap the throttle wide open and return to idle. Note and record the peak reading. This reading should be higher than the idle reading. Snap throttle pressure _____

5. Record your readings for running and snap compression for all cylinders. The running compression reading should be approximately 50 psi to 75 psi, and snap compression should be about 80 percent of cranking compression.

Running Pressure

1. _____   5. _____

2. _____   6. _____

3. _____   7. _____

4. _____   8. _____

Snap Pressure

1. _____   5. _____

2. _____   6. _____

3. _____   7. _____

4. _____   8. _____

Worn cam lobes and weak or broken valve springs can cause low running compression. Higher-than normal readings, over 80 percent of cranking compression pressure can be caused by a restricted exhaust system.

6. Based on your testing, what conclusions can you make about the engine?

7. Remove any tools and equipment and clean up the work area.

Instructor's check _____

Lab Worksheet 23-7

Name _____ Station _____ Date _____

Relative Compression Test

ASE Education Foundation Correlation

This lab worksheet addresses the following **MLR** task:

8.A.4. Perform cylinder cranking and running compression tests; document results. **(P-2)**

Description of Vehicle

Year _____ Make _____ Model _____

Engine _____ AT MT CVT PSD (circle which applies)

Procedure

1. Disable the fuel and ignition systems. Explain how this is accomplished. _____

    _____

    Warning! Hybrid vehicles require special procedures for conducting compression tests. Do not try to perform this test on a hybrid without following the manufacturer's precautions and procedures.

2. Connect a current probe to a lab scope.

3. Set up the current probe and lab scope to display cranking current draw.

    a. Current probe setting _____

    b. Lab scope voltage range _____

    c. Lab scope time base _____

4. Crank the engine and capture the cranking current waveform.

5. Describe or draw the waveform pattern. _____

    _____

    _____

6. Determine the peak and lowest amperages during cranking.

   a. Peak amperage_____

   b. Lowest amperage _____

7. Based on your test results, what is the condition of the engine? _____
   _____
   _____

8. Remove any tools and equipment and clean up the work area.

                              Instructor's check _____

Lab Worksheet 23-8

Name _____ Station _____ Date _____

Transducer Compression Test

ASE Education Foundation Correlation

This lab worksheet addresses the following **MLR** task:

8.A.4. Perform cylinder cranking and running compression tests; document results. **(P-2)**

Description of Vehicle

Year _____ Make _____ Model _____

Engine _____ AT MT CVT PSD (circle which applies)

Procedure

1. Connect and calibrate the pressure transducer to the lab scope.

   a. Pressure range selected _____

   b. Time base selected _____

2. Remove the spark plug from the cylinder being tested. Install the pressure transducer in the spark plug hole.

3. Crank the engine and obtain a compression waveform.
   Maximum pressure recorded _____

4. Start the engine and obtain several cycles of cylinder compression.

   a. Maximum cylinder running pressure _____

   b. Minimum cylinder running pressure _____

5. Snap the throttle wide open and return to idle. Note and record the peak reading.
   Snap throttle pressure _____

6. Based on your testing, what is the condition of the cylinder? _____

   _____

   _____

7. Remove any tools and equipment and clean up the work area.

   Instructor's check _____

Lab Worksheet 23-9

Name _____ Station _____ Date _____

Cylinder Leakage Test

ASE Education Foundation Correlation

This lab worksheet addresses the following **MLR** task:

8.A.5. Perform cylinder leakage test; document results. **(P-2)**

Description of Vehicle

Year _____ Make _____ Model _____

Engine _____ AT MT CVT PSD (circle which applies)

Procedure

1. With the engine cool, remove the spark plug from the cylinder to be tested.

2. Connect the cylinder leak tester to the shop air and zero out the tester. If this step is not performed, the test 'results will not be accurate.

   Instructor's check _____

3. Make sure the Schrader valve is removed from the cylinder hose and install the hose into the spark plug hole.

4. Set the engine so that the cylinder being tested is at TDC compression. You may need to prevent the crankshaft from spinning by using a socket and ratchet to hold the crank pulley bolt.

   Instructor's check _____

5. Connect the leakage tester hose to the cylinder hose and note the reading on the gauge.

   Gauge reading _____

6. Is the reading excessive?   Yes _____   No _____

   If yes, try to determine where the air is escaping from. Note your results. _____

   _____

7. Based on your testing, what is the condition of the cylinder? _____

_____

_____

8. Remove any tools and equipment and clean up the work area.

Instructor's check _____

Lab Worksheet 23-10

Name _____ Station _____ Date _____

Inspect Exhaust

ASE Education Foundation Correlation

This lab worksheet addresses the following **MLR** tasks:

8.C.3. Inspect integrity of the exhaust manifold, exhaust pipes, muffler(s), catalytic converter(s), resonator(s), tail pipe(s), and heat shields; determine necessary action. **(P-1)**

8.C.4. Inspect condition of exhaust system hangers, brackets, clamps, and heat shields; determine necessary action. **(P-1)**

Description of Vehicle

Year _____ Make _____ Model _____

Engine _____ AT MT CVT PSD (circle which applies)

Procedure

1. Refer to the manufacturer's service information for specific procedures to inspect the exhaust system. Summarize the procedures. _____

   _____

   _____

   _____

2. Inspect the exhaust system for evidence of damage, missing or broken components, or aftermarket parts. Note your findings. _____

   _____

   _____

3. Use a rubber mallet to lightly tap on the exhaust system to check for loose parts. Note the results of this test.

   _____

   _____

4. Check that all heat shields are in place and securely mounted. Note your findings.

_____

_____

5. Have an assistant start the engine while you listen for exhaust leaks. Be careful not to be burned on any hot exhaust system parts. Note your findings. _____

_____

_____

6. Based on your inspection, what is the condition of the exhaust system and what actions are necessary?

_____

_____

_____

7. Remove any tools and equipment and clean up the work area.

Instructor's check _____

CHAPTER 24

Engine Performance Service

Review Questions

1. Until the 1980s, engine and transmission systems were, in nearly all cars and trucks, nearly 100 percent _____ operated.

2. Even with on-board computers to manage engine operation, some items, such as _____ filters, spark plugs, and other normal wear items, still require periodic _____ and replacement.

3. The _____ of gasoline's vapors makes working around gasoline hazardous.

4. List five safety precautions for working with gasoline.

   a. _____

   b. _____

   c. _____

   d. _____

   e. _____

5. _____ fuel filters have either plastic or steel shells and contain a _____ filter element.

6. *Technician A* says all modern vehicles have a serviceable inline fuel filter. *Technician B* says some vehicles place the fuel filter in the fuel tank and it is routinely replaced as part of a maintenance program. Who is correct?

   a. Technician A

   b. Technician B

   c. Both A and B

   d. Neither A nor B

7. Before you begin to replace a fuel filter, the fuel system must be _____ to prevent fuel spraying from an open line.

8. *Technician A* says special tools may be required to replace a fuel filter. *Technician B* says fuel filters are directional and must be installed a certain way. Who is correct?

   a. Technician A

   b. Technician B

   c. Both A and B

   d. Neither A nor B

9. Some vehicles are equipped with an airflow _____ gauge to help determine when the _____ filter should be replaced.

10. Oil in the air cleaner housing or ducts often indicates a problem with the _____ system.

11. Identify the parts of the spark plug shown in Figure 24-1.

   A. _____   B. _____

   C. _____   D. _____

   E. _____   F. _____

   G. _____

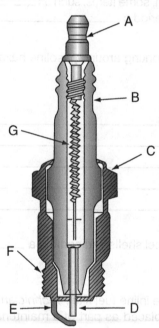

Figure 24-1

12. Which is *not* a typical spark plug replacement interval?

   a. 30,000 miles

   b. 60,000 miles

   c. 100,000 miles

   d. All the above

13. Describe why it is important to let the engine cool before removing the spark plugs.

_____

_____

_____

14. Explain what you should do if a spark plug is very difficult to remove. _____

_____

_____

_____

15. List three things spark plugs should be inspected for. _____

_____

_____

16. *Technician A* says antiseize should be used on all new spark plugs. *Technician B* says all spark plugs should be torqued to specifications. Who is correct?

   a. Technician A

   b. Technician B

   c. Both A and B

   d. Neither A nor B

17. Spark plug gap specifications are typically between _____ and _____.

18. Caution must be used when you are checking and gapping plugs with _____ and _____ electrodes as these are easily damaged.

19. Explain how to properly reinstall and tighten a spark plug. _____

_____

_____

_____

20. Explain how the PCV valve is used to reduce emissions. _____

_____

_____

_____

21. If left unchecked, the crankcase vapors will build up _____, causing _____ leaks.

22. *Technician A* says at idle, the PCV valve is fully open and flowing vapors back into the intake manifold. *Technician B* says the PCV valve is closed under low vacuum conditions. Who is correct?

    a. Technician A

    b. Technician B

    c. Both A and B

    d. Neither A nor B

23. Describe how to test a PCV valve. _____

    _____

    _____

    _____

24. In addition to checking the PCV valve and hose, check the _____ air hose for cracks, rotting, and tight connections.

25. OBD I systems were used until _____.

26. OBD I arose from the need to reduce exhaust _____.

27. List three items standardized by OBD II.

    a. _____

    b. _____

    c. _____

28. A major requirement of OBD II is to illuminate the malfunction indicator light if emissions increase _____ percent over the federal test parameter for that vehicle.

29. List eight systems monitored by OBD II.

    a. _____

    b. _____

    c. _____

    d. _____

    e. _____

    f. _____

    g. _____

    h. _____

30. Describe the differences between global and enhanced data. _____

    _____

    _____

    _____

31. A(n) _____ is a test run by the ECM on components and systems to test their operation and to determine that certain operating conditions have been met.

32. Explain how passive, active, and intrusive tests differ from each other. _____

_____

_____

_____

33. *Technician A* says the CCM is used to monitor sensor information and determine if wiring faults are present. *Technician B* says the CCM is used to actively test component operation. Who is correct?

   a. Technician A

   b. Technician B

   c. Both A and B

   d. Neither A nor B

34. Which component is used to determine engine misfire?

   a. Throttle position sensor

   b. Crankshaft position sensor

   c. Engine coolant temperature sensor

   d. All the above

35. A misfire code P0304 indicates a misfire on cylinder number _____.

36. Type A misfires must be detected within _____ to _____ crankshaft revolutions.

37. Explain what is meant by closed-loop operation. _____

_____

_____

_____

38. Fuel control is represented by _____ term and _____ term fuel trim on the scan tool.

39. *Technician A* says an STFT and LTFT reading of 18 percent indicates that the engine is running lean and the computer is adding fuel to the mixture. *Technician B* says a vacuum leak may cause the lean condition that the computer is adding fuel to compensate for. Who is correct?

   a. Technician A

   b. Technician B

   c. Both A and B

   d. Neither A nor B

40. Explain why some monitors, such as EVAP and catalyst efficiency, are noncontinuous monitors.

_____

_____

_____

41. *Technician A* says the catalyst efficiency monitor is dependent upon the oxygen sensor operation. *Technician B* says a faulty oxygen sensor may cause the catalyst monitor not to complete. Who is correct?

    a. Technician A

    b. Technician B

    c. Both A and B

    d. Neither A nor B

42. Describe the purpose of the EVAP system. _____

_____

_____

_____

43. The EGR system is used to reduce _____ temperatures.

44. A drive cycle is a series of trips designed to allow all _____ to run.

45. Explain enable criteria as related to a trip. _____

_____

_____

46. Regarding the OBD II diagnostic trouble code P0301: *Technician A* says a P code indicates a powertrain code. *Technician B* says that P0 indicates the code is generic. Who is correct?

    a. Technician A

    b. Technician B

    c. Both A and B

    d. Neither A nor B

47. A _____ code appears when the ECM has detected a problem but is waiting for a second trip to confirm the code.

48. If the VIN information is incorrectly entered into a scan tool during setup, which of the following may result?

    a. No communication

    b. Unusual data

    c. Damage to the scan tool

    d. Both a and b

Lab Activity 24-1

Name _____ Date _____ Instructor _____

Year _____ Make _____ Model _____

Engine _____ Scan tool _____

On-Board Computer Communication Checks

1. Connect a scan tool to the DLC. Select Global OBD II and begin communication. List the communication protocol displayed on the scan tool. _____

2. List the number of ECUs the scan tools detects. _____

3. Check and record any stored DTCs. _____

4. Exit Global OBD II and enter enhanced OBD (EOBD) mode. List the systems available on the scan tool menu. (e.g., powertrain, ABS, airbag, etc.) _____

   _____

   _____

5. Does the scan tool enable you to perform a network test?   Yes _____   No _____

   If yes, perform the network test and note the results. _____

   _____

6. If the scan tool does not have a network test, try to communicate with each system shown from the main menu. Note your results. _____

   _____

   Based on your inspection, is the network operating correctly?   Yes _____   No _____

Lab Worksheet 24-1

Name _____ Station _____ Date _____

RR Gaskets

ASE Education Foundation Correlation

This lab worksheet addresses the following **MLR** task:

1.A.4. Install engine covers using gaskets, seals, and sealers as required. **(P-1)**

Description of Vehicle

Year _____ Make _____ Model _____

Engine _____ AT MT CVT PSD (circle which applies)

Procedure

1. Using service information, locate and describe the procedures for installing this component.

   _____

   _____

2. Clean the component so that none of the old gasket material remains and so it is clean and dry.

   Instructor's check _____

3. Which is being used?   Gasket   Seal   Gasket maker (RTV or similar)

4. If using a gasket, determine the correct orientation for the gasket and secure the gasket in place. This may require threading bolts through the bolt holes in the gasket. Install the component while making sure the gasket remains aligned and in place.

   Instructor's check _____

5. If installing a seal, determine which direction the seal faces when installed. Install the seal. This may require the use of a special seal installation tool.

   Instructor's check _____

6. If using a form-in-place gasket maker, read the directions on the container. Form a bead of the gasket material on the component. Summarize the correct use of this type of sealer.

Instructor's check _____

7. Once the component is installed, install the remaining fasteners and torque them to specifications. Note: Some components require the fasteners to be torqued in a specific sequence.

Torque specification _____

Torque procedure _____

Instructor's check _____

8. Once the component is installed, clean the work area and verify your repair and that there are no leaks.

Instructor's check _____

9. Once finished, put away all tools and equipment and clean up the work area.

Instructor's check _____

Lab Worksheet 24-2

Name _____ Station _____ Date _____

RR TBelt

ASE Education Foundation Correlation

This lab worksheet addresses the following **MLR** task:

1.A.5. Verify engine mechanical timing. **(P-2)**

Description of Vehicle

Year _____ Make _____ Model _____

Engine _____ AT MT CVT PSD (circle which applies)

Procedure

1. Using service information, locate and print the procedures for replacing the timing belt.

2. List the components that must be removed to access the timing covers. _____
   _____
   _____

3. Describe how the crank and camshafts are to be set to remove the timing belt. _____
   _____

4. How will you be able to verify the engine is properly timed to remove the belt? _____
   _____

5. Set the crank and camshafts as described in the service information.

   Instructor's check _____

6. If not already removed, remove the timing covers and note the condition of the timing belt and its related components.
   _____
   _____

7. Are the indications of oil or coolant leaks in the timing belt area?   Yes _____   No _____

8. Install any sprocket holding devices as needed. Release the tension on the belt as described in the service information and remove the belt. Replace any timing pulleys as necessary and torque fasteners to specifications.

Instructor's check _____

9. Install the new belt and apply tension as described in the service information.

Instructor's check _____

10. Rotate the crankshaft two complete turns and recheck the timing sprocket alignment marks.

    Do all timing marks align properly?   Yes _____   No _____

    If no, repeat steps 9 and 10.

Instructor's check _____

11. Once the crank and cam timing is verified, recheck belt tension.

Instructor's check _____

12. Reassemble the engine's front end components. Start the engine and verify proper engine operation.

Instructor's check _____

13. Once finished, put away all tools and equipment and clean up the work area.

Instructor's check _____

Lab Worksheet 24-3

Name _____ Station _____ Date _____

Adjust Valves

ASE Education Foundation Correlation

This lab worksheet addresses the following **MLR** task:

1.B.1. Adjust valves (mechanical or hydraulic lifters). **(P-3)**

Description of Vehicle

Year _____ Make _____ Model _____

Engine _____ AT MT CVT PSD (circle which applies)

Procedure

1. Using the service information, locate the procedures to adjust the valves for this vehicle.

   Instructor's check _____

2. Determine the valve clearance specifications.

   Intake valve clearance _____

   Exhaust valve clearance _____

3. List the order in which the valves are adjusted. _____

   _____

4. Using the service information, begin the adjustment procedure. Record the measured clearances before adjustment. Record both exhaust and intake valve readings for each valve Set each valve's clearance to specifications. Tighten all fasteners to specifications.

   Cylinder 1   Exhaust _____   Intake _____

   Cylinder 2   Exhaust _____   Intake _____

   Cylinder 3   Exhaust _____   Intake _____

Cylinder 4   Exhaust _____   Intake _____

Cylinder 5   Exhaust _____   Intake _____

Cylinder 6   Exhaust _____   Intake _____

Cylinder 7   Exhaust _____   Intake _____

Cylinder 8   Exhaust _____   Intake _____

Rocker arm nut torque specification _____

Instructor's check _____

5. Once the adjustment is complete, reinstall all covers and torque fasteners to specifications.

Instructor's check _____

6. Start the engine and listen for any noise from the valvetrain. Tapping noises can be caused by excessive valve clearance. Engine noise: _____

Instructor's check _____

7. Once finished, put away all tools and equipment and clean up the work area.

Instructor's check _____

Lab Worksheet 24-4

Name _____ Station _____ Date _____

Inspect Spark Plugs

ASE Education Foundation Correlation

This lab worksheet addresses the following **MLR** task:

8.A.7. Remove and replace spark plugs; inspect secondary ignition components for wear and damage. **(P-1)**

Description of Vehicle

Year _____ Make _____ Model _____

Engine _____ Number of cylinders _____

Procedure

1. Locate and record the manufacturer's service procedures for removing and installing spark plugs.
   _____
   _____
   _____
   _____

2. Locate the engine's firing order and note it here. _____

3. Locate the engine's cylinder arrangement and draw it here.

4. Before beginning, make sure the engine is cool to the touch. Do not attempt to remove spark plugs from a hot engine.

5. Carefully remove the spark plug wire or coil from the spark plug. Note any difficulties.
   _____
   _____

6. Note the size of the spark plug socket needed to remove the spark plug. _____

7. Install the spark plug socket onto the spark plug. Using a ratchet, break the plug loose and remove it from the engine. Note any difficulties. _____

8. Examine the spark plug. Note the condition of the following:

   a. Plug wire terminal _____

   b. Insulation _____

   c. Steel shell and hex _____

   d. Threads _____

   e. Center and ground electrodes _____

   f. Color of the electrodes and insulator _____

9. Using a spark plug gapping tool, carefully measure and record the plug gap.

   Measured gap _____   Gap specification _____

10. Determine if antiseize is used on the spark plug thread.   Yes _____   No _____

11. Locate and record the spark plug torque specification. _____

12. Carefully thread the spark plug into the cylinder head. Hand-tighten the plug until it seats against the head and then torque the plug to specifications.

   Instructor's check _____

13. Reinstall the spark plug wire or coil.

   Instructor's check _____

14. Inspect the components of the secondary ignition system. Check for damaged connectors, cracks, high-voltage burn through, and corrosion. Note your findings.

   Ignition cables _____

   Ignition coil(s) _____

   Distributor cap _____

   Ignition rotor _____

15. Remove any tools and equipment and clean up the work area.

   Instructor's check _____

Lab Worksheet 24-5

Name _____ Station _____ Date _____

Retrieve DTCs

ASE Education Foundation Correlation

This lab worksheet addresses the following **MLR** task:

8.B.1. Retrieve and record diagnostic trouble codes (DTC), OBD monitor status, and freeze frame data; clear codes when applicable. **(P-1)**

Description of Vehicle

Year _____ Make _____ Model _____

Engine _____ AT MT CVT PSD (circle which applies)

Scan tool used _____

Procedure

1. Connect the scan tool and turn the ignition on. Does the scan tool power up correctly?

   Yes _____   No _____

   If no, what does this indicate? _____

2. Does the scan tool support a module or network status test?

   Yes _____   No _____

   If yes, what is the test called? _____

3. If the scan tool supports a module/network test, perform the test and record the results.

   _____

   _____

   _____

4. If the scan tool does not support module and network tests, attempt to communicate with modules known to be present on the vehicle. Record your results. _____

_____

_____

5. Based on your testing, are all modules active on the network?   Yes _____ No _____

6. Do any of the modules have stored DTCs?   Yes _____ No _____

   If yes, record the DTCs. _____

7. Select an accessory system such as lighting for testing. System selected _____

8. List any bidirectional (active) command functions available in the system selected.

_____

_____

9. How can this tool capability be used to diagnose and repair a problem with the vehicle? _____

_____

_____

10. Remove any tools and equipment and clean up the work area.

                    Instructor's check _____

Lab Worksheet 24-6

Name _____ Station _____ Date _____

Check Freeze Frame Data

ASE Education Foundation Correlation

This lab worksheet addresses the following **MLR** task:

8.D.1. Retrieve and record diagnostic trouble codes (DTC), OBD monitor status, and freeze frame data; clear codes when applicable. **(P-1)**

Description of Vehicle

Year _____ Make _____ Model _____

Engine _____ AT MT CVT PSD (circle which applies)

Procedure

1. Connect a scan tool to the vehicle's DLC. Scan tool used _____

2. Navigate to the DTC menu and record any stored DTCs.

   Current DTCs _____

   Pending DTCs _____

   History DTCs _____

3. Select a DTC from the freeze frame data. DTC selected _____

4. Record the following information from the freeze frame record:

   Mileage or starts since first/last fail _____

   Pass/fail counter _____

   Rpm at time of fault _____

Fuel trims at time of fault _____

Coolant temperature at time of fault _____

MAP/MAF at time of fault _____

5. Based on the freeze frame data, what can you determine to be the possible cause of the DTC? _____
_____

_____

6. Remove any tools and equipment and clean up the work area.

Instructor's check _____

Lab Worksheet 24-7

Name _____ Station _____ Date _____

Monitor Status

ASE Education Foundation Correlation

This lab worksheet addresses the following **MLR** task:

8.B.2. Describe the use of the OBD monitors for repair verification. **(P-1)**

Description of Vehicle

Year _____ Make _____ Model _____

Engine _____ AT MT CVT PSD (circle which applies)

Procedure

1. Connect a scan tool to the vehicle's DLC. Scan tool used _____

2. Enter Global OBD II and begin communication.

3. Navigate to the OBD monitor status and record the condition of each of the following:

Misfire _____   Fuel control _____

Component (CCM) _____   Catalyst _____

HO_2S _____ O_2 Heater _____

EVAP _____   EGR _____

Secondary air _____   PCV _____

Thermostat _____   Air conditioning _____

4. Are any monitors incomplete or failed?   Yes _____   No _____

   If incomplete, what may be the cause? _____

   _____

   If failed, what may be the cause? _____

   _____

5. How can checking emission monitor status be used to verify a repair is complete? _____

   _____

   _____

   _____

6. Based on your inspection, what conclusion can you make about the OBD system? _____

   _____

   _____

7. Remove any tools and equipment and clean up the work area.

                              Instructor's check _____

Lab Worksheet 24-8

Name _____ Station _____ Date _____

Relieve pressure

ASE Education Foundation Correlation

This lab worksheet addresses the following **MLR** task:

8.C.1. Replace fuel filter(s) where applicable. **(P-2)**

Description of Vehicle

Year _____ Make _____ Model _____

Engine _____ AT MT CVT PSD (circle which applies)

Procedure

1. Locate and record the manufacturer's procedures for relieving fuel pressure. _____
   _____
   _____

   If the manufacturer does not specify a method to relieve fuel pressure, select one of the following as the correct procedure:

   a. Install a fuel pressure gauge on the pressure test port and release the fuel from the gauge into a suitable container.
   b. Remove the fuel pump fuse, start the engine, and idle until it stalls.
   c. Remove the fuel pump relay, start the engine, and idle until it stalls.
   d. Disconnect the fuel pump connection, start the engine, and idle until it stalls.

2. Method selected _____

   Before any work is performed on the fuel system, any remaining fuel pressure must be relieved.

3. Once work is complete, restore the fuel system to its normal condition.

4. Turn the key on and check for fuel leaks before starting the engine.

5. Start the engine and verify proper fuel system operation.

Instructor's check _____

6. Remove any tools and equipment and clean up the work area.

Instructor's check _____

Lab Worksheet 24-9

Name _____ Station _____ Date _____

Replace Fuel Filters

ASE Education Foundation Correlation

This lab worksheet addresses the following **MLR** task:

8.C.1. Replace fuel filter(s) where applicable. **(P-2)**

Description of Vehicle

Year _____ Make _____ Model _____

Engine _____ AT MT CVT PSD (circle which applies)

WARNING: Do not smoke, carry lighted tobacco or have an open flame of any type when working on or near any fuel-related component. Highly flammable mixtures are always present and may be ignited. Do not carry personal electronic devices such as cell phones, pagers or audio equipment of any type when working on or near any fuel-related component. Highly flammable mixtures are always present and may be ignited. Before working on or disconnecting any of the fuel tubes or fuel system components, relieve the fuel system pressure to prevent accidental spraying of fuel. Fuel in the fuel system remains under high pressure, even when the engine is not running. When handling fuel, always observe fuel handling precautions and be prepared in the event of fuel spillage. Spilled fuel may be ignited by hot vehicle components or other ignition sources.

Procedure

1. Locate and record the manufacturer's procedures for relieving fuel pressure. _____

    _____

    _____

2. Location of fuel filter. _____

3. Once fuel pressure is relieved, place a shop pan under the filter to catch any remaining fuel that spills out once the fuel lines are removed.

    Instructor's check _____

4. Remove the fuel lines and the filter.

    Instructor's check _____

5. Install the new filter and attach the fuel lines.

Instructor's check _____

6. Verify the filter and fuel lines are secured properly.

Instructor's check _____

7. Enable the fuel system. Turn the ignition ON and check for fuel leaks. *Do not* start the engine if a leak is present.

Instructor's check _____

8. If there are no leaks, start the engine and let it idle for approximately one minute to remove any air from the fuel supply system.

Instructor's check _____

9. Remove any tools and equipment and clean up the work area.

Instructor's check _____

Lab Worksheet 24-10

Name _____ Station _____ Date _____

Inspect Air Filter

ASE Education Foundation Correlation

This lab worksheet addresses the following **MLR** tasks:

8.C.1. Replace fuel filter(s) where applicable. **(P-2)**

8.C.2. Inspect, service, or replace air filters, filter housings, and intake duct work. **(P-1)**

Description of Vehicle

Year _____ Make _____ Model _____

Engine _____ AT MT CVT PSD (circle which applies)

Procedure

1. Refer to the manufacturer's service information for specific procedures to inspect the air induction system and air filter. Summarize the procedures. _____

   _____

   _____

2. Inspect the air induction system for evidence of damage, missing components, or aftermarket parts. Note your findings. _____

   _____

   _____

3. Remove the air filter from the air filter housing. Inspect the housing for dirt, debris, and damage. Note your findings. _____

   _____

4. Inspect the air filter element for excessive dirt or debris buildup, damage, and incorrect application or installation concerns. Note your findings. _____

   _____

   _____

5. Based on your inspection, what is/are the necessary action(s)? _____
    _____
    _____

                                            Instructor's check _____

6. Reinstall or replace the air filter as necessary.

7. Remove any tools and equipment and clean up the work area.

                                            Instructor's check _____

Lab Worksheet 24-11

Name _____ Station _____ Date _____

Inspect PCV

ASE Education Foundation Correlation

This lab worksheet addresses the following **MLR** task:

8.D.1. Inspect, test, and service positive crankcase ventilation (PCV) filter/breather, valve, tubes, orifices, and hoses; perform necessary action. **(P-2)**

Description of Vehicle

Year _____ Make _____ Model _____

Engine _____ AT MT CVT PSD (circle which applies)

Procedure

1. Using the service information, locate and record the manufacturer's procedures for inspecting the PCV valve and system. _____

   _____

   _____

   _____

2. Find and note the location of the PCV valve. _____

   _____

3. With the engine running, carefully remove the PCV valve. Place your thumb over the valve's opening and note the reaction of the valve. _____

4. Reinstall the valve into the engine and inspect the vacuum hose to the valve. Note the condition of the vacuum hose. _____

5. Locate the fresh air hose to the engine for the PCV system. Inspect the air hose and note its condition.

   _____

   _____

6. Based on your inspection, what is the condition of the PCV system? _____
_____
_____

7. Remove any tools and equipment and clean up the work area.

Instructor's check _____

CHAPTER 25

Drivetrains and Transmissions

Review Questions

1. The _____ allows the engine to run at its most efficient speeds so that power is not wasted.

2. Describe in your own words about the three types of drivetrain configurations._____
   _____
   _____

3. List three types of transmissions used in modern vehicles.
   a. _____
   b. _____
   c. _____

4. As on-board computer systems became standard, the _____ transmission also began to be controlled by a powertrain control module or PCM.

5. List three examples of inputs to the PCM used for transmission control.
   a. _____
   b. _____
   c. _____

6. What are three benefits of having the computer control transmission operation?
   a. _____
   b. _____
   c. _____

7. List six major components or systems of a modern automatic transmission.
   a. _____
   b. _____
   c. _____
   d. _____
   e. _____
   f. _____

641

8. The _____ converter is used to join the engine's crankshaft to the input shaft of an automatic transmission.

9. Label the components shown in Figure 25-1.

   A. _____   B. _____

   C. _____

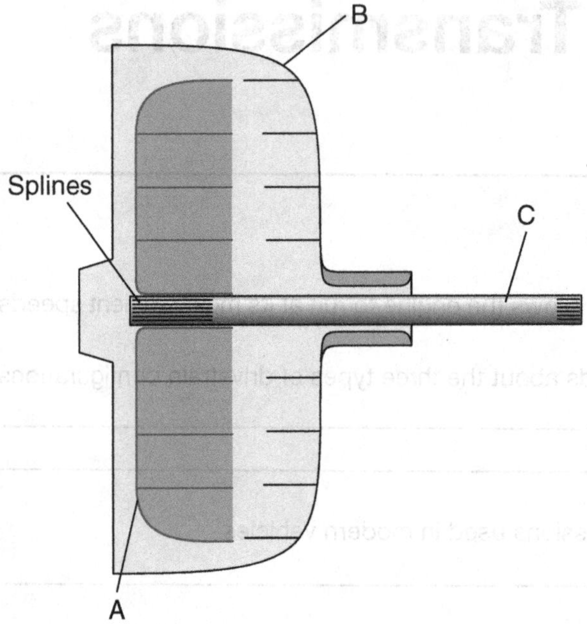

Figure 25-1

10. The torque converter uses _____ to transmit power.

11. Which torque converter component is used to drive the input shaft?

    a. Impeller

    b. Stator

    c. Turbine

    d. Shell

12. Which device is used to achieve optimal torque converter efficiency at cruising speed?

    a. Stator

    b. Lock-up clutch

    c. Oil pump

    d. None of the above

13. The torque converter is only about _____ efficient until the lock-up clutch applies.

14. The rear of the torque converter drives the front transmission _____ pump.

15. Label the parts of the planetary gearset shown in Figure 25-2.

    A. _____   B. _____

    C. _____   D. _____

    E. _____

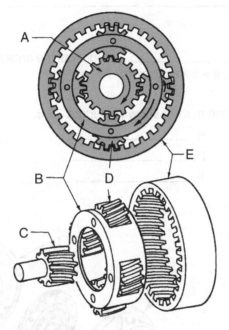

Figure 25-2

16. Describe the operation of a clutch pack in an automatic transmission. _____
    _____
    _____

17. Before the use of electrically operated transmission shift _____ became standard, all of the shifting functions were controlled by the _____ system.

18. A _____ solenoid is used to open and close a hydraulic passage.

19. Describe how a CVT transmits power from the engine to the drive wheels. _____
    _____
    _____

20. *Technician A* says CVT transmissions are used to maintain efficient engine speeds. *Technician B* says some CVT vehicles are programed to have shift points and manual shifting modes. Who is correct?

    a. Technician A

    b. Technician B

    c. Both A and B

    d. Neither A nor B

21. Some hybrid vehicles use a _____ device instead of a conventional transmission.

22. With a manual transmission, a _____ connects the engine to the transmission and transmits the power from the crankshaft to the transmission.

23. Label the components of the clutch system shown in Figure 25-3.

    A. _____   B. _____

    C. _____   D. _____

    E. _____   F. _____

    G. _____

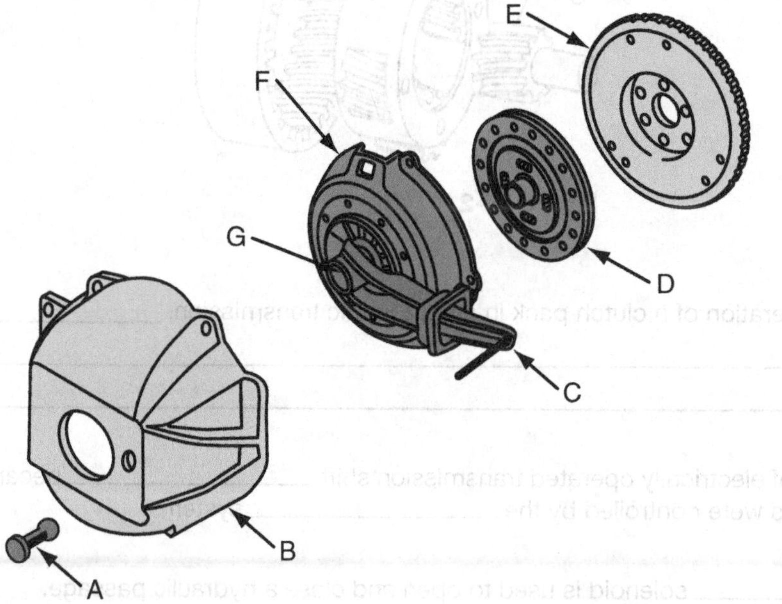

Figure 25-3

24. Explain the purpose and operation of the pressure plate. _____

    _____

    _____

    _____

    _____

25. Describe two types of clutch release bearings. _____
_____
_____
_____

26. _____ gear ratios increase torque and allow for quick acceleration.

27. _____ gear ratios allow lower engine speed and economy but with low torque output.

28. List the five major components of a manual transmission.
    a. _____
    b. _____
    c. _____
    d. _____
    e. _____

29. Define *gear ratio*. _____
_____
_____

30. Match the following gear ratios with the gear number.

    0.8:1   Second gear
    1:1   Third gear
    1.3:1   Overdrive (fifth gear)
    2:1   First gear
    3.5:1   Fourth gear

31. Explain the purpose and operation of a synchronizer. _____
_____
_____

32. The shift linkage connects the gear shifter to the _____ shift valve.

33. List five common transmission services.
    a. _____
    b. _____
    c. _____
    d. _____
    e. _____

34. To accurately check the transmission fluid on a vehicle with a dipstick, the fluid usually must be at _____ temperature.

35. *Technician A* says most modern automatic transmissions use the same type of transmission fluid. *Technician B* says most automatic transmissions have specific fluid requirements. Who is correct?

    a. Technician A

    b. Technician B

    c. Both A and B

    d. Neither A nor B

36. *Technician A* says some automatic transmissions have a drain plug and do not have a filter that requires periodic replacement. *Technician B* says some automatic transmissions do not have a dipstick for checking fluid level. Who is correct?

    a. Technician A

    b. Technician B

    c. Both A and B

    d. Neither A nor B

37. List four problems that can be caused by a faulty transmission range switch.

    a. _____

    b. _____

    c. _____

    d. _____

38. List four problems that can be caused by worn transmission mounts.

    a. _____

    b. _____

    c. _____

    d. _____

39. Describe how to check for worn powertrain mounts. _____

    _____

    _____

    _____

40. When a leak occurs from a manual transmission, check for a plugged _____, which can allow pressure to increase inside the transmission.

41. Explain how to bleed a hydraulic clutch system. _____

    _____

    _____

    _____

42. _____ differentials generally require a specific type of lubricant be used, which contain an additive of special friction modifiers.

43. When an outer CV joint is worn excessively, it will cause a loud _____ sound during turns.

44. Describe how to replace a FWD axle shaft. _____

_____

_____

_____

45. Describe how to remove and replace a U-joint. _____

_____

_____

_____

46. An older 4WD vehicle may have _____ locking hubs.

47. What is indicated by a clicking or ratcheting sound from a 4WD front hub?

_____

_____

_____

Activities

A gear is a circular component that transmits rotational force to another gear or component. Gears have teeth so that they can mesh with other gears without slipping. In Figure 25-4, the driving or input gear has a radius of 1 foot and the driven or output gear has a radius of 2 feet. In this example, the driving gear is turning with 10 pounds of force, which results in 20 pounds of force provided by the driven gear. The output force is increased due to the larger size of the driven gear as each tooth acts as a lever. Because the driven gear has twice as many teeth as the driving gear, the driven gear will turn at one-half of the speed of the driving gear. This equals a gear ratio of 2:1, with two turns of the driving gear to one turn of the driven gear.

Determine the gear ratios for the gear arrangements shown in Figure 25-5.

Teeth of input gear apply
10 pound of force to output gear teeth.

10 lb-ft. Torque

2 ft.

1 ft.

20 lb-ft. Torque

Input   Output

10 lb-ft. ÷ 1 ft. = 10 lb × 2 ft. = 20 lb-ft.

Figure 25-4

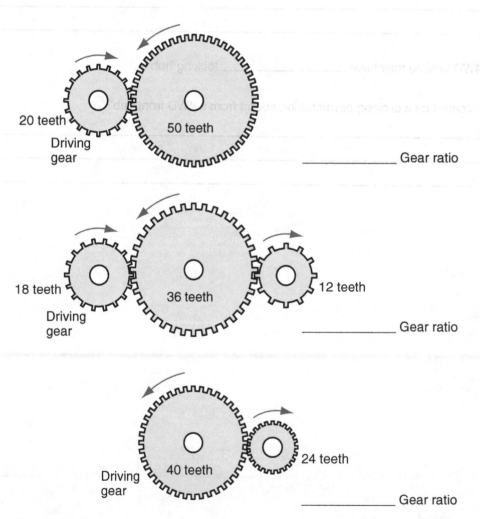

20 teeth
Driving gear

50 teeth

_____ Gear ratio

18 teeth
Driving gear

36 teeth

12 teeth

_____ Gear ratio

Driving gear

40 teeth

24 teeth

_____ Gear ratio

Figure 25-5

Lab Worksheet 25-1

Name _____ Station _____ Date _____

Trans Fluid ID

ASE Education Foundation Correlation

This lab worksheet addresses the following **MLR** task:

2.A.1. Research vehicle service information including fluid type, vehicle service history, service precautions, and technical service bulletins. **(P-1)**

Description of Vehicle

Year _____ Make _____ Model _____

Engine _____ AT Model _____

Procedure

1. Determine the correct fluid type for the transmission/transaxle and the service refill quantity.

   Fluid type specification _____

   Service refill capacity _____

2. Determine if any service has been done on the transmission for this vehicle. Note your findings.

   _____

   _____

3. Using the service information, locate any technical service bulletins (TSBs) related to the transmission for this vehicle. Record the TSB numbers and publication dates. _____

   _____

   _____

4. Using the service information, locate and record any service precautions for working on the transmission on this vehicle.

   _____

   _____

Lab Worksheet 25-2

Name _____ Station _____ Date _____

Transmission ID

ASE Education Foundation Correlation

This lab worksheet addresses the following **MLR** task:

2.A.1. Research vehicle service information including fluid type, vehicle service history, service precautions, and technical service bulletins. **(P-1)**

Description of Vehicle

Year _____ Make _____ Model _____

Engine _____ AT MT CVT PSD (circle which applies)

Procedure

1. Safely raise and support the vehicle.

2. Locate the transmission and look for any identification tags.

    Is there an ID tag?   Yes _____   No _____

    a. If yes, what information is contained on the tag? _____
    _____

    b. If no, are there any markings that indicate the transmission manufacturer or other information?

    Yes _____   No _____

    Manufacturer _____

3. If no ID tag is found, make a drawing of the transmission pan and note the location of the pan bolts.

4. Using the drawing of the transmission pan, identify the pan and transmission using the service information for the vehicle. Transmission model _____

5. Remove any tools and equipment and clean up the work area.

Instructor's check _____

Lab Worksheet 25-3

Name _____ Station _____ Date _____

Transmission ID

Transmission Fluid Check

This lab worksheet addresses the following **MLR** tasks:

2.A.2. Check fluid level in a transmission or a transaxle equipped with a dip-stick. **(P-1)**

2.A.3. Check fluid level in a transmission or a transaxle not equipped with a dip-stick. **(P-1)**

2.A.4. Check transmission fluid condition; check for leaks. **(P-2)**

Description of Vehicle

Year _____ Make _____ Model _____

Engine _____ AT MT CVT PSD (circle which applies)

Procedure

1. Refer to the owner's manual or service information for the recommended fluid type and record it here.
   _____

2. Does the vehicle have a transmission dipstick?   Yes _____   No _____

3. Locate and record the manufacturer's method for checking the fluid level.
   _____
   _____

   a. Does the transmission fluid need to be within a certain temperature range?

      Yes _____   No _____   Range _____

   b. If yes, operate the vehicle until the fluid is within the specified temperature range.

   c. How will you determine the temperature of the fluid? _____
      _____

4. For vehicles with a dipstick: Remove the dipstick and draw a picture of the end of the stick where the fluid level is checked. Include instructions about how to check the fluid and what type of fluid should be used.

     a. Reinstall and again remove the dipstick; note the fluid level as indicated.

       _____

     b. Describe the appearance of the fluid. _____

       _____

5. For vehicles without a dipstick, determine fluid level and note your results.

    _____

6. Based on your inspection, summarize the condition of the transmission fluid.

    _____

    _____

7. Inspect all visible portions of the transmission/transaxle for signs for fluid leaks. Note your findings.

    _____

    _____

8. Safely raise and support the vehicle. Check the underside of transmission/transaxle for signs of fluid leaks. Note your findings. _____

    _____

9. Inspect axle or driveshaft seals for signs of fluid leaks. Note your findings. _____

    _____

10. Inspect the transmission cooler lines and hoses for signs of fluid leaks. Note your findings.

     _____

     _____

11. Inspect around the shift linkage, electrical connections, and dipstick tube for signs of fluid leaks. Note your findings. _____

     _____

12. Based on your inspection, what are the necessary actions? _____

     _____

     _____

13. Remove any tools and equipment and clean up the work area.

                                           Instructor's check _____

Lab Worksheet 25-4

Name _____ Station _____ Date _____

Drive Train Configuration

ASE Education Foundation Correlation

This lab worksheet addresses the following **MLR** task:

2.A.5. Identify drive train components and configuration. **(P-1)**

Description of Vehicle

Year _____ Make _____ Model _____

Engine _____ AT MT CVT PSD (circle which applies)

Procedure

1. Safely raise and support the vehicle.

2. Determine the drive train configuration.   FWD   RWD   AWD/4WD

   a. How can you determine the drive train configuration?

   _____

3. Locate the following:

   A. FWD transaxle   _____   E. FWD drive shafts   _____

   B. RWD transmission   _____   F. RWD rear prop shaft   _____

   C. Transfer case/center differential _____   G. 4WD front prop shaft   _____

   D. Front differential   _____   H. Rear differential   _____

                                              Instructor's check _____

Lab Worksheet 25-5

Name _____ Station _____ Date _____

Drive Train Configuration

ASE Education Foundation Correlation

This lab worksheet addresses the following **MLR** task:

2.A.5. Identify drive train components and configuration. **(P-1)**

Description of Vehicle

Year _____ Make _____ Model _____

Engine _____ AT Model _____

Procedure

1. Refer to the manufacturer's service information regarding how to inspect, adjust, or replace the shift linkage and related sensors or switches. Summarize the procedures. _____

    _____

    _____

2. Perform a visual inspection of the shift linkage and transmission range sensor. Note your findings.

    _____

    _____

3. Install a scan tool and locate the transmission data. With the key on engine off (KOEO), move the gear selector from Park through each gear range and note the gear position data on the scan tool. Does the gear position data match the movement of the gear selector?   YES   NO

4. If the gear position data does not match the movement of the gear selector, adjust the linkage following the manufacturer's service procedures.

    Instructor's check _____

5. If the sensor or switch cannot be adjusted, it may need to be replaced. Describe the procedure to replace the sensor or switch. _____

    _____

6. Once the linkage has been adjusted or the sensor replaced, confirm proper operation of the shift linkage and transmission range sensor or switch.

Instructor's check _____

7. Once finished, put away all tools and equipment and clean up the work area.

Instructor's check _____

Lab Worksheet 25-6

Name _____ Station _____ Date _____

Inspect Transmission Fluid Leaks

ASE Education Foundation Correlation

This lab worksheet addresses the following **MLR** task:

2.B.2. Inspect for leakage at external seals, gaskets, and bushings. **(P-1)**

Description of Vehicle

Year _____ Make _____ Model _____

Engine _____ AT Model _____

Procedure

1. Safely raise and support the vehicle.

2. Inspect the following transmission/transaxle areas for signs of fluid leaks. Note your findings.

Fluid leak at bell housing	YES	NO
Fluid leak at output shaft seal	YES	NO
Fluid leak at axle shaft seals	YES	NO
Fluid leak from pan gasket	YES	NO
Fluid leak from side pan gasket	YES	NO
Fluid leak at cooler line connections	YES	NO
Fluid leak at dipstick tube	YES	NO
Fluid leak at vent	YES	NO

3. If fluid is leaking from the output shaft or axle shaft seals, what other component may be faulty and causing the leak? _____

4. Based on your inspection, what is/are the necessary action(s)? _____

5. Once finished, put away all tools and equipment and clean up the work area.

Instructor's check _____

Lab Worksheet 25-7

Name _____ Station _____ Date _____

Inspect and Replace Power Train Mounts

ASE Education Foundation Correlation

This lab worksheet addresses the following **MLR** task:

2.B.3. Inspect, replace and/or align power train mounts. **(P-2)**

Description of Vehicle

Year _____ Make _____ Model _____

Engine _____ AT MT CVT PSD (circle which applies)

Procedure

1. Locate the manufacturer's service information for inspecting the powertrain mounts. Summarize the inspection process. _____

_____

_____

2. Following the manufacturer's inspection procedures, inspect the mounts and record your findings.

_____

_____

_____

3. Describe what types of problems worn or loose powertrain mounts can cause. _____

_____

_____

_____

4. To replace a powertrain mount, remove the fasteners holding the two sections together. Next, carefully raise the engine or transmission/transaxle to remove the mount.

Instructor's check _____

5. Remove the fasteners holding the mount in place and remove the mount.

   Instructor's check _____

6. Compare the replacement part with the original and ensure the replacement part is correct. Install the new mount and torque the fasteners to specifications.

   Torque specifications _____

   Instructor's check _____

7. Align the mount and carefully lower the engine or transmission/transaxle into place. Install and tighten the remaining mount fasteners to specifications.

   Instructor's check _____

8. Start the engine and apply the brakes. Check for excessive vibration, movement, or noise from the mount(s) when shifting the transmission from Park to Reverse and Drive.

   Instructor's check _____

9. Once finished, put away all tools and equipment and clean up the work area.

   Instructor's check _____

Lab Worksheet 25-8

Name _____ Station _____ Date _____

Automatic Transmission Fluid Service

ASE Education Foundation Correlation

This lab worksheet addresses the following **MLR** task:

2.B.4. Drain and replace fluid and filter(s); use proper fluid type per manufacturer specification. **(P-1)**

Description of Vehicle

Year _____ Make _____ Model _____

Engine _____ AT MT CVT PSD (circle which applies)

Procedure

1. Determine the correct fluid type for the transmission/transaxle and the service refill quantity.

   Fluid type specification _____

   Service refill capacity _____

2. If the transmission/transaxle does not have a serviceable filter, locate the fluid drain plug, place a fluid drain-pan under the transmission/transaxle, and remove the drain plug. Allow the fluid to drain completely and reinstall the drain plug and torque to specifications. Skip to step 13.

   Drain plug torque specification _____

3. If the transmission/transaxle has a serviceable filter, you will need to remove the pan to access the filter. NOTE – some transmissions have a threaded canister filter similar to an engine oil filter. These are replaced without removing the transmission/transaxle pan. Some transmission have both an internal and an external filter. This transmission has which type(s) of filter? _____

4. Place a large drain pan under the transmission/transaxle and loosen the pan bolts approximately ¼ turn. Continue to loosen all the pan bolts except the bolts on one corner of the pan. This will allow the pan to tilt down slowly and the fluid to drain out without make as much of a mess.

   Instructor's check _____

5. Once most of the fluid has drained, support the pan and remove the remaining corner bolts. Carefully lower the pan into the drain pan. Make sure you have all the pan bolts. Clean off the pan bolts and inspect the threads for damage. _____

    Instructor's check _____

6. Remove the old gasket from the pan and the transmission/transaxle case as necessary. There should not be any old gasket material stuck to either the pan or the case.

    Instructor's check _____

7. Remove the filter from the valve body and place it inside the drain pan.

    Instructor's check _____

8. If applicable, clean where the filter attaches to the valve body and replace any seals as required to attach the new filter. Install the new filter making sure it is properly attached to the valve body.

    Instructor's check _____

9. Inspect the fluid in the transmission/transaxle pan for debris. A small amount is normal as the friction part wear. There should not be any metal shavings. Note your findings. _____

    _____

10. Clean and dry the transmission/transaxle pan. Install the new gasket to the pan using the pan bolts to keep the gasket in place.

    Instructor's check _____

11. Align the pan with the transmission/transaxle case and place the pan in position. Begin threading the pan bolts back into the case. Make sure each bolt starts and threads smoothly into the holes in the case.

    Instructor's check _____

12. Following the manufacturer's service information, begin tightening the pan bolts as directed and torque them to specifications.

    Pan bolt tightening sequence _____

    Pan bolt torque specification _____

13. Install a funnel and begin refilling the transmission/transaxle with the correct fluid.

    Instructor's check _____

14. Once refilled, start the engine and check for leaks from the transmission/transaxle pan.

Instructor's check _____

15. Run the transmission to operating temperature and make sure the fluid is filled per the manufacturer's recommendations.

   Correct fluid temperature range to check level. _____

Instructor's check _____

16. Once finished, put away all tools and equipment and clean up the work area.

Instructor's check _____

14. Once refilled, start the engine and check for leaks from the transmission/transaxle pan.

_____ Instructor's check

15. Run the transmission to operating temperature and make sure the fluid is filled per the manufacturer's recommendations.

_____ Correct fluid temperature range to check level

_____ Instructor's check

16. Once finished, put away all tools and equipment and clean up the work area.

_____ Instructor's check

Lab Worksheet 25-9

Name _____ Station _____ Date _____

Describe CVT

ASE Education Foundation Correlation

This lab worksheet addresses the following **MLR** task:

2.C.1. Describe the operational characteristics of a continuously variable transmission (CVT). **(P-3)**

Description of Vehicle

Year _____ Make _____ Model _____

Engine _____ Transmission Model _____

Procedure

Refer to the manufacturer's service information regarding the description and operation for the transmission for this vehicle.

1. Describe how a CVT is different from a planetary gear based automatic transmission and from a constant mesh manual transmission. _____

   _____

   _____

   _____

2. Describe the control system for the CVT. _____

   _____

   _____

3. Does this CVT use a starting clutch?   YES   NO

   If YES, describe the starting clutch operation _____

   _____

   _____

4. Explain how the manufacturers provide shift feel for CVTs. _____

   _____

Lab Worksheet 25-10

Name _____ Station _____ Date _____

Describe HEV Drive Train

ASE Education Foundation Correlation

This lab worksheet addresses the following **MLR** task:

2.C.2. Describe the operational characteristics of a hybrid vehicle drive train. **(P-3)**

Description of Vehicle

Year _____ Make _____ Model _____

Engine _____ Transmission Type _____

Procedure

Refer to the manufacturer's service information regarding the description and operation for the transmission or hybrid drive system for this vehicle.

1. Describe how the transmission or hybrid drive system is different from a traditional automatic transmission and from a manual transmission. _____

   _____

   _____

   _____

2. Describe the control system for the system used by this vehicle. _____

   _____

   _____

3. Explain how the electric motors are used in the transmission or hybrid drive system for this vehicle.

   _____

   _____

Lab Worksheet 25-10

Name _____ Station _____ Date _____

Describe HEV Drive Train

ASE Education Foundation Correlation

This lab worksheet addresses the following MLR task:

2.C.2. Describe the operational characteristics of a hybrid vehicle drive train. (P-3)

Description of Vehicle

Year _____ Make _____ Model _____

Engine _____ Transmission Type _____

Procedure

Refer to the manufacturer's service information regarding the description and operation for the transmission or hybrid drive system for this vehicle.

1. Describe how the transmission or hybrid drive system is different from a traditional automatic transmission and from a manual transmission.

_____

_____

_____

2. Describe the control system for the system used by this vehicle.

_____

_____

_____

3. Explain how the electric motors are used in the transmission or hybrid drive system for this vehicle.

_____

_____

Lab Worksheet 25-11

Name _____ Station _____ Date _____

TITLE

ASE Education Foundation Correlation

This lab worksheet addresses the following **MLR** task:

3.A.1. Research vehicle service information including fluid type, vehicle service history, service precautions, and technical service bulletins. **(P-1)**

Description of Vehicle

Year _____ Make _____ Model _____

Engine _____ MT Model _____

Procedure

1. Determine the correct fluid type for the transmission/transaxle and the service refill quantity.

   Fluid type specification _____

   Service refill capacity _____

2. Determine if any service has been done on the transmission for this vehicle. Note your findings.

3. Using the service information, locate any technical service bulletins (TSBs) related to the transmission for this vehicle. Record the TSB numbers and publication dates. _____

   _____

   _____

4. Using the service information, locate and record any service precautions for working on the transmission on this vehicle. _____

   _____

   _____

Lab Worksheet 25-12

Name _____ Station _____ Date _____

Drain and Refill Manual Transmission Fluid

ASE Education Foundation Correlation

This lab worksheet addresses the following **MLR** tasks:

3.A.2. Drain and refill manual transmission/transaxle and final drive unit; use proper fluid type per manufacturer specification. **(P-1)**

3.A.3. Check fluid condition; check for leaks. **(P-2)**

Description of Vehicle

Year _____ Make _____ Model _____

Engine _____ Transmission _____

Procedure

1. Refer to the owner's manual or service information for the recommended fluid type and record it here.

   _____

2. Locate and record the manufacturer's method for checking the fluid level. _____

   _____

   _____

3. Remove the fluid fill plug and note the level of fluid in the transmission.

   Full _____   Low _____

   a. How did you determine the fluid level?_____

   _____

4. Use a fluid extraction device to remove a small amount of fluid. Inspect the fluid and compare its appearance to new fluid. Describe the condition of the fluid. _____

   _____

5. Check the condition of the fill plug and gasket; note any concerns. _____

_____

6. To drain the fluid, place a shop pan or fluid waste container under the transmission/transaxle drain. Remove the drain plug. Note the condition of the fluid compared to the sample you checked at the fill plug.

_____

_____

7. Once fully drained, reinstall the drain plug and torque it to specifications.

   Torque specs. _____

8. Visually inspect the transmission for evidence of fluid loss. Note your findings. _____

_____

9. Refill the transmission/transaxle with the fluid specified by the manufacturer.

   Fluid capacity _____

10. Once the transmission/transaxle is full, reinstall the fill plug and torque it to specifications.

    Torque specs. _____

11. Remove any tools and equipment and clean up the work area.

Instructor's check _____

Lab Worksheet 25-13

Name _____ Station _____ Date _____

Drive Train Configuration

ASE Education Foundation Correlation

This lab worksheet addresses the following **MLR** task:

3.A.4. Identify manual drive train and axle components and configuration. **(P-1)**

Description of Vehicle

Year _____ Make _____ Model _____

Engine _____ AT MT CVT PSD (circle which applies)

Procedure

1. Safely raise and support the vehicle.

2. Determine the drive train configuration. FWD RWD AWD/4WD

   a. How can you determine the drive train configuration? _____

   _____

3. Locate the following:

   A. FWD transaxle   _____   E. FWD drive shafts   _____

   B. RWD transmission   _____   F. RWD rear prop shaft   _____

   C. Transfer case/center differential   _____   G. 4WD front prop shaft   _____

   D. Front differential   _____   H Rear differential   _____

   Instructor's check _____

Lab Worksheet 25-14

Name _____ Station _____ Date _____

Check Clutch Fluid and Inspect for Leaks

ASE Education Foundation Correlation

This lab worksheet addresses the following **MLR** tasks:

3.B.1. Check and adjust clutch master cylinder fluid level; use proper fluid type per manufacturer specification. **(P-1)**

3.B.2. Check for hydraulic system leaks. **(P-1)**

Description of Vehicle

Year _____ Make _____ Model _____

Engine _____ MT Model _____

Procedure

1. Determine the correct fluid type for the hydraulic clutch system.

    Fluid type _____

2. If the clutch master cylinder reservoir is opaque plastic, examine the fluid level inside the master cylinder and note your findings. _____

    _____

3. If the fluid level cannot be seen through the master cylinder reservoir, clean the reservoir cap and area around the cap. Remove the cap and note the fluid level. _____

    _____

4. What could cause the fluid level to be low? Be specific. _____

    _____

    _____

5. If the fluid level is low, top off the reservoir with the correct fluid.

Instructor's check _____

6. Inspect the master cylinder body, rear seal where the pushrod enters the cylinder, and the steel line for leaks. Note your findings. _____

_____

7. Inspect the clutch slave cylinder body and steel line connection for leaks. Note your findings.

_____

_____

8. Based on your inspection, summarize the condition of the hydraulic clutch system.

_____

_____

9. Once finished, put away all tools and equipment and clean up the work area.

Instructor's check _____

Lab Worksheet 25-15

Name _____ Station _____ Date _____

Describe Electronically-Controlled Manual Transmission

ASE Education Foundation Correlation

This lab worksheet addresses the following **MLR** task:

3.C.1. Describe the operational characteristics of an electronically-controlled manual transmission/transaxle. **(P-2)**

Description of Vehicle

Year _____ Make _____ Model _____

Engine _____ Transmission/transaxle model _____

Procedure

Refer to the manufacturer's service information for the description and operation of the transmission for this vehicle.

1. Describe how the transmission differs from a traditional automatic transmission and from a manual transmission. _____

   _____

   _____

   _____

2. Describe the control system for the system used by this vehicle. _____

   _____

   _____

3. Explain how the gears are shifted in this transmission. _____

   _____

   _____

4. What are some advantages of this type of transmission? _____

   _____

Lab Worksheet 25-16

Name _____ Station _____ Date _____

CV Shaft Diagnosis and Removal

ASE Education Foundation Correlation

This lab worksheet addresses the following **MLR** task:

3.D.2. Inspect, service, and/or replace shafts, yokes, boots, and universal/CV joints. **(P-2)**

Description of Vehicle

Year _____ Make _____ Model _____

Engine _____ AT MT CVT PSD (circle which applies)

Procedure

1. Worn constant velocity joints are a common problem with FWD vehicles. Describe the types of problems and customer concerns associated with worn inner and outer CV joints. _____

   _____

   _____

   _____

2. Describe how to diagnose worn inner and outer CV joints. _____

   _____

   _____

   Once a worn or damaged CV joint is diagnosed, you will typically remove the axle shaft to replace either the faulty joint or the entire axle with a remanufactured unit.

3. Locate the manufacturer's service information for inspecting and replacing a FWD drive shaft. Summarize the inspection process. _____

   _____

   _____

4. Safely raise and support the vehicle so the front suspension is unloaded.

5. Remove the wheel and tire. Remove the axle nut. NOTE – the axle nuts are typically very tight. Have an assistant hold the brake pedal down to prevent the axle from spinning while removing the nut.

6. Loosen and remove either the strut-to-knuckle bolts or the lower ball joint to separate the knuckle from the lower control arm.

Instructor's check _____

7. Attempt to push the outer CV joint and shaft from the hub. Depending on the vehicle, a hub puller may be required to push the axle shaft out of the hub.

Instructor's check _____

8. Depending on the vehicle, the inner CV joint may be bolted to a flange at the transaxle or it may be held in place with a retaining ring. Describe how to remove the axle from the transaxle based on the service information. _____

_____

9. Remove the axle shaft from the vehicle.

Instructor's check _____

Lab Worksheet 25-17

Name _____ Station _____ Date _____

CV Joint Replacement

ASE Education Foundation Correlation

This lab worksheet addresses the following **MLR** task:

3.D.2. Inspect, service, and/or replace shafts, yokes, boots, and universal/CV joints. **(P-2)**

Description of Vehicle

Year _____ Make _____ Model _____

Engine _____ AT MT CVT PSD (circle which applies)

Procedure

1. With the axle shaft removed, refer to the manufacturer's service information to determine how to remove the faulty CV joint. Summarize the procedure. _____

   _____

   _____

2. Remove the clamps from the CV boot then remove the CV boot.

   Instructor's check _____

3. Remove the CV joint from the axle shaft. Clean the axle shaft and replace the retaining ring if necessary Inspect the replacement joint and ensure it matches the original joint.

   Instructor's check _____

4. Following the instructions provided with the new CV boot kit, slide the new boot and small clamp onto the axle shaft. Next, install the CV joint onto the axle shaft.

   SERVICE NOTE – you may need to pack the new joint with the supplied grease before installing it onto the shaft. Make sure the joint is fully seated and locked to the shaft.

   Instructor's check _____

5. Make sure the CV boot is installed properly so the inboard clamp will secure the boot in the correct location. Secure the small clamp using the appropriate tool.

Instructor's check _____

6. Squeeze any remaining grease into the CV boot. Set the outboard side of the boot into position onto the joint and place the clamp around the boot. When properly aligned, secure the boot clamp using the appropriate tool.

Instructor's check _____

7. Clean any remaining grease from the boot and shaft.

Instructor's check _____

8. Once finished, put away all tools and equipment and clean up the work area.

Instructor's check _____

Lab Worksheet 25-18

Name _____ Station _____ Date _____

CV Joint Replacement

ASE Education Foundation Correlation

This lab worksheet addresses the following **MLR** task:

3.D.2. Inspect, service, and/or replace shafts, yokes, boots, and universal/CV joints. **(P-2)**

Description of Vehicle

Year _____ Make _____ Model _____

Engine _____ AT MT CVT PSD (circle which applies)

Procedure

1. If installing a replacement axle shaft, make sure the new axle is the same as the original. Inspect the splines, axle length, and ensure ABS rings are installed if applicable.

2. Position the shaft to align with the transaxle and attach the inboard joint in/to the transaxle. If the joint is bolted to the transaxle, locate the torque specifications and toque the bolts to specs.

    Torque specifications _____

    Instructor's check _____

3. Slide the outboard CV shaft into the hub. Depending on the fit, you may need to use the axle nut to help draw the joint through the hub.

    Instructor's check _____

4. Reattach the strut-to-knuckle bolts or the lower ball joint and toque the fasteners to specifications.

    Torque specifications _____

    Instructor's check _____

5. Torque the axle nut to specifications. Do not use power tools to seat the axle nut. Overtightening the axle nut can damage the wheel bearing and cause premature bearing failure.

    Torque specifications _____

    Instructor's check _____

6. What problems could result from improperly torqueing the axle nut? _____

_____

_____

7. Some axle nuts are staked once they are torqued. This prevents the nut from working loose. Is the axle nut on this vehicle staked?   YES   NO

8. If NO, what prevents the axle nut from working loose?_____

_____

_____

9. Once finished, put away all tools and equipment and clean up the work area.

Instructor's check _____

Lab Worksheet 25-19

Name _____ Station _____ Date _____

Inspect Front Wheel Bearings 4WD

ASE Education Foundation Correlation

This lab worksheet addresses the following **MLR** task:

3.D.3. Inspect locking hubs. **(P-3)**

Description of Vehicle

Year _____ Make _____ Model _____

Engine _____ AT MT CVT PSD (circle which applies)

Procedure

1. Locate and record the manufacturer's inspection procedures for the front-wheel bearings and locking hubs for this vehicle. _____

   _____

   _____

2. Safely raise and support the vehicle with the front suspension hanging.

   Instructor's check _____

3. Check for play in the front bearings by rocking the tire up and down and side to side. Note your findings.

   _____

   _____

4. Spin the tire and listen for noise from the wheel bearing. You may need to use a stethoscope to hear bearing noise. Note your findings. _____

   _____

   _____

5. Engage the 4WD system so that the front hubs are locked. Spin a front tire and note of the other front tire rotates at the same time. Note your findings. _____

_____

6. Based on your inspection, what is the condition of the front-wheel bearings and locking hubs?

_____

_____

7. Describe two possible causes for the locking hubs not to operate correctly. _____

_____

_____

8. Remove any tools and equipment and clean up the work area.

Instructor's check _____

Lab Worksheet 25-20

Name _____ Station _____ Date _____

Inspect Locking Hubs

ASE Education Foundation Correlation

This lab worksheet addresses the following **MLR** task:

3.D.3. Inspect locking hubs. **(P-3)**

Description of Vehicle

Year _____ Make _____ Model _____

Engine _____ AT MT CVT PSD (circle which applies)

Procedure

Safely raise and support the vehicle.

1. Grasp the wheel and tire at the 12 and 6 o'clock positions and shake the tire in and out to check for loose wheel bearings. Repeat while grasping the tire at the 3 and 9 o'clock positions. Was any looseness noted?

   YES   NO

2. If YES, repeat the test while checking to see if the play is from the wheel bearing or other component of the steering or suspension systems. Note your findings. _____

   _____

3. The tire should spin without requiring excessive effort. Does it spin easily?   YES   NO

4. Move the locking hub selector to LOCK. Spin the tire again. Does spinning the tire also spin the front driveshaft?

   YES   NO

5. If the front driveshaft does not spin when spinning the tire, what does this indicate?

   _____

6. Switch the locking hub selector back to UNLOCK and spin the tire. Does the tire spin freely?

   YES   NO

7. Based on your inspection, what can you determine about the locking hubs? _____

_____

_____

8. Once finished, put away all tools and equipment and clean up the work area.

Instructor's check _____

Lab Worksheet 25-21

Name _____ Station _____ Date _____

Check Differential and Transfer Case for Leaks

ASE Education Foundation Correlation

This lab worksheet addresses the following **MLR** task:

3.D.4. Check for leaks at drive assembly and transfer case seals; check vents; check fluid level; use proper fluid type per manufacturer specification. **(P-2)**

Description of Vehicle

Year _____ Make _____ Model _____

Engine _____ AT MT CVT PSD (circle which applies)

Procedure

Safely raise and support the vehicle.

1. Inspect the final drive assembly for leaks at the axle seals and at the drive shaft seal. Describe your findings.

_____

_____

_____

2. A blocked differential vent can allow pressure to build and cause fluid leaks. Locate the vent. The vent may be a cap with a loose top or may be attached to the end of a rubber hose. Describe the vent on this vehicle.

_____

_____

3. Inspect the vent for blockage or damage. Note your findings. _____

_____

4. Locate the fluid check/fill plug. Remove the plug and note the fluid level. Typically, the fluid level should be up to the bottom of the fill plug hole.

Fluid level _____

5. If the fluid level is low, what type of fluid is specified by the manufacturer to refill the unit?

   Fluid type _____

6. Inspect the transfer case assembly for leaks. Describe your findings. _____

   _____

   _____

7. Locate the fluid check/fill plug on the transfer case. Remove the plug and note the fluid level. Typically, the fluid level should be up to the bottom of the fill plug hole.

   Fluid level _____

8. If the fluid level is low, what type of fluid is specified by the manufacturer to refill the unit?

   Fluid type _____

9. Summarize the results of your inspection. _____

   _____

   _____

10. Once finished, put away all tools and equipment and clean up the work area.

                                                          Instructor's check _____

Lab Worksheet 25-22

Name _____ Station _____ Date _____

Check Differential Case

ASE Education Foundation Correlation

This lab worksheet addresses the following **MLR** task:

3.E.1. Clean and inspect differential case; check for leaks; inspect housing vent. **(P-1)**

Description of Vehicle

Year _____ Make _____ Model _____

Engine _____ AT MT CVT PSD (circle which applies)

Procedure

Safely raise and support the vehicle.

1. Inspect the front drive assembly for leaks at the axle seals and at the drive shaft seal. Describe your findings.

   _____

   _____

   _____

2. A blocked differential vent can allow pressure to build and cause fluid leaks. Locate the vent. The vent may be a cap with a loose top or may be attached to the end of a rubber hose. Describe the vent on this vehicle.

   _____

   _____

3. Inspect the vent for blockage or damage. Note your findings. _____

   _____

4. Locate the fluid check/fill plug. Remove the plug and note the fluid level. Typically, the fluid level should be up to the bottom of the fill plug hole.

   Fluid level _____

5. Summarize the results of your inspection. _____

_____

_____

6. Once finished, put away all tools and equipment and clean up the work area.

Instructor's check _____

Lab Worksheet 25-23

Name _____ Station _____ Date _____

Inspect Differential Fluid

ASE Education Foundation Correlation

This lab worksheet addresses the following **MLR** task:

3.E.2. Check and adjust differential case fluid level; use proper fluid type per manufacturer specification. **(P-1)**

Description of Vehicle

Year _____ Make _____ Model _____

Engine _____ AT MT CVT PSD (circle which applies)

Procedure

1. Refer to the owner's manual or service information for the recommended fluid type and record it here. If inspecting a 4WD vehicle with two differentials, note the fluid type for each. _____

   _____

2. Locate and record the manufacturer's method for checking the fluid level. _____

   _____

   _____

3. Remove the fluid fill plug and note the level of fluid in the transmission.

   Full _____   Low _____

   How did you determine the fluid level? _____

   _____

4. Reinstall the fill plug and torque it to specifications (if applicable).

   Torque spec. _____

5. Visually inspect the differential for evidence of fluid loss. Note your findings.

_____

6. Remove any tools and equipment and clean up the work area.

Instructor's check _____

Lab Worksheet 25-24

Name _____ Station _____ Date _____

Drain and Refill Differential

ASE Education Foundation Correlation

This lab worksheet addresses the following **MLR** task:

3.E.3. Drain and refill differential housing. **(P-1)**

Description of Vehicle

Year _____ Make _____ Model _____

Engine _____ Differential type _____

Procedure

1. Using the service information, determine the correct fluid type and capacity.

   Differential fluid type _____

   Fluid capacity _____

   NOTE – Vehicles with limited slip differentials (LSDs) typically require a special additive to the differential lubricant. Is the differential you are working on a LSD?   YES   NO

   If YES, what type of fluid or additive is required? _____

2. Some differentials have a drain plug but most require removing a cover to drain the fluid. Some shops use a fluid extractor to remove the old fluid. How is the fluid drained from this differential? _____

   _____

3. If the cover is removed, you will need to reseal the cover with a gasket or by using a form-in-place gasket maker. Which will you use on this differential? _____

4. Before draining the fluid, remove the differential check/fill plug. This is to be sure fluid can be added after the differential has been drained.

   Instructor's check _____

5. Drain the fluid from the differential. If a drain plug was removed, reinstall the drain plug and torque to specifications.

   Torque specifications _____

                                          Instructor's check _____

6. If the cover was removed, remove the old gasket material from the cover and the differential housing. Clean the housing cover in preparation of applying a new gasket.

                                          Instructor's check _____

7. Install the new gasket to the housing cover. If form-in-place gasket maker is being used, follow the directions on product and apply a bead of gasket material to the cover.

                                          Instructor's check _____

8. Install the cover on the differential housing. Make sure the bolt holes align and begin threading the cover bolts into the case. *Caution: Do not tighten any of the bolts until all the bolts are started and threaded into the case.*

9. Once all the bolts are in place, torque the bolts to specifications.

   Torque specifications _____

                                          Instructor's check _____

10. Refill the differential with the specified lubricant(s) until the fluid is up to the check/full plug hole. Reinstall the check/fill plug and torque to specifications if applicable.

    Torque specifications _____

                                          Instructor's check _____

11. Once finished, put away all tools and equipment and clean up the work area.

                                          Instructor's check _____

Heating and Air Conditioning

CHAPTER 26

Review Questions

1. The HVAC system is responsible for maintaining engine _____ and for passenger _____.

2. Describe the components that make up the HVAC system. _____

_____

_____

3. Explain the main functions of the cooling system. _____

_____

_____

4. List eight components of the cooling system.

   a. _____

   b. _____

   c. _____

   d. _____

   e. _____

   f. _____

   g. _____

   h. _____

5. During combustion, temperatures inside the engine can reach _____°F.

6. _____ is the medium by which heat transfer takes place between the engine and the surrounding air.

7. *Technician A* says coolant color is the best method of determining which coolant is used in a vehicle. *Technician B* says all ethylene glycol-based coolants are interchangeable and can be mixed. Who is correct?

   a. Technician A

   b. Technician B

   c. Both A and B

   d. Neither A nor B

8. Hybrid electric vehicles often use _____ different coolants.

9. The water pump may be driven by the _____ belt or by a(n) _____ drive belt.

10. The _____ transfers the heat of the coolant to the outside air.

11. Explain why pressure is allowed to build up in the cooling system during operation.

    _____

    _____

    _____

12. Some cooling systems use a _____ tank, also called a _____ bottle as the fill point for the system.

13. The radiator fan may be driven by a belt, an _____ motor, or hydraulically from the _____ system.

14. Describe the purpose of the fan clutch. _____

    _____

    _____

    _____

15. Explain the purpose of the fan shroud. _____

    _____

    _____

    _____

    _____

16. The thermostat uses a _____ pellet to move a piston.

17. Thermostats are usually fully open by _____°F.

18. A mini radiator, called the _____, is used to supply heat to the passenger compartment.

19. Before performing any work on the cooling system, you first must determine what _____ of coolant is required.

20. Describe how to perform a cooling system pressure test. _____

   _____

   _____

   _____

   _____

21. *Technician A* says a faulty cooling system pressure cap can cause the engine to overheat. *Technician B* says all cooling system caps should hold at least 15 psi of pressure. Who is correct?

   a. Technician A

   b. Technician B

   c. Both A and B

   d. Neither A nor B

22. Describe how to test for coolant leaking into the combustion chamber. _____

   _____

   _____

   _____

   _____

23. Explain how to inspect cooling system hoses and hose clamps. _____

   _____

   _____

   _____

   _____

24. *Technician A* says a defective thermostat can cause the engine to overheat. *Technician B* says a defective thermostat can cause the engine to run too cold. Who is correct?

   a. Technician A

   b. Technician B

   c. Both A and B

   d. Neither A nor B

25. A faulty thermostat can cause the _____ light to come on and cause a diagnostic _____ to be set in the computer.

26. _____ the cooling system removes the old coolant and circulates clean water through the system to remove rust and corrosion.

27. To remove trapped _____ from the cooling system, many systems are equipped with a bleed valve.

28. The _____ system works with both the heating and air condition systems to route air flow for the passenger compartment.

29. List three components found in the HVAC case.

    a. _____

    b. _____

    c. _____

30. _____ air filters are often mounted near the heater core/evaporator housing and are used to trap and prevent _____, _____, and other contaminants from entering the passenger compartment.

31. Explain how stepped resistors are used to control blower motor speeds. _____

    _____

    _____

32. List three methods of controlling the operation of the blend door.

    a. _____

    b. _____

    c. _____

33. List six components of the air conditioning system.

    a. _____

    b. _____

    c. _____

    d. _____

    e. _____

    f. _____

34. *Technician A* says refrigerant has a very high boiling point, so it can absorb a lot of heat. *Technician B* says refrigerant is kept under pressure to prevent it from boiling. Who is correct?

    a. Technician A

    b. Technician B

    c. Both A and B

    d. Neither A nor B

35. The _____ separates the high- and low-pressure sides of the AC system.

36. On hybrid vehicles, the compressor may be driven by a _____ or by high-voltage _____.

37. The high-voltage cables and components used in hybrid _____ vehicles are bright in color to easily distinguish them from other components.

38. An accumulator is used on the _____ pressure side of the system.

39. The component that is used to remove heat from the refrigerant is the _____.

40. To reduce the pressure of the refrigerant entering the evaporator, either an _____ tube or _____ valve is used.

41. Explain four precautions for working with the AC system.

    a. _____

    b. _____

    c. _____

    d. _____

42. Describe what problems the AC drive belt should be inspected for. _____

    _____

    _____

    _____

    _____

43. Cabin air filters may be located near the _____ air inlet in the cowl or in the dash, near the blower motor.

44. During AC operation, _____ forms on the evaporator, which can cause mildew and odor from the HVAC system.

Activities

1. Label the components of the cooling system shown in Figure 26-1.

   A. _____   B. _____

   C. _____   D. _____

   E. _____   F. _____

   G. _____

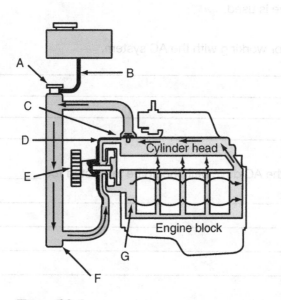

Figure 26-1

2. Label the high and low sections of the AC system shown in Figure 26-2.

BASIC SYSTEM SCHEMATIC

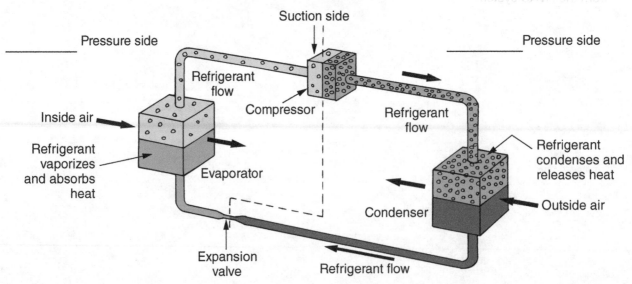

Figure 26-2

3. Label the components of the AC system shown in Figure 26-3.

A. _____   B. _____

C. _____   D. _____

E. _____

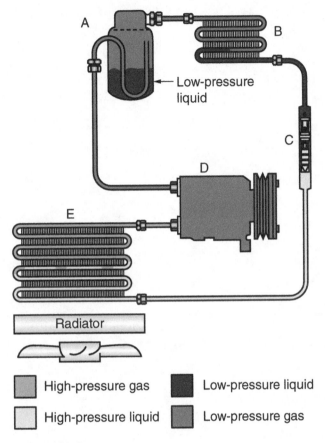

Figure 26-3

8. Label the components of the AC system shown in Figure 26-2.

A. _____   B. _____

C. _____   D. _____

E. _____

Low-pressure liquid

Radiator

High-pressure gas   ▨ Low-pressure liquid

High-pressure liquid   ▨ Low-pressure gas

Figure 26-3

Lab Activity 26-1

Name _____ Date _____ Instructor _____

Determine Correct Coolants

Using a selection of vehicles, determine the correct coolant for each. To determine the correct coolant, you may need to check the owner's manual, service information, and the Internet.

1. Year _____ Make _____ Model _____

    Engine _____

    Recommended coolant _____

    Cooling system capacity _____

2. Year _____ Make _____ Model _____

    Engine _____

    Recommended coolant _____

    Cooling system capacity _____

3. Year _____ Make _____ Model _____

    Engine _____

    Recommended coolant _____

    Cooling system capacity _____

Lab Worksheet 26-1

Name _____ Station _____ Date _____

Visual Leak Inspection

ASE Education Foundation Correlation

This lab worksheet addresses the following **MLR** task:

1.C.1. Perform cooling system pressure and dye tests to identify leaks; check coolant condition and level; inspect and test radiator, pressure cap, coolant recovery tank, and heater core and gallery plugs; determine necessary action. **(P-1)**

Description of Vehicle

Year _____ Make _____ Model _____

Engine _____ AT MT CVT PSD (circle which applies)

Procedure

Inspect the following components and note your findings. _____

1. Pressure cap rating _____ psi   Bar   kPa (circle one)

2. Check for leaks around pressure cap. Note your findings. _____

3. Inspect the radiator for signs of leaks. Note your findings. _____

4. Inspect hoses for leaks, cracks or bulges. Note your findings. _____

5. Inspect radiator fan for damage. Note your findings. _____

6. Inspect for coolant leaks around the thermostat and hoses. Note your findings. _____
   _____

7. Start the engine and set the heater controls to high heat output.

8. Does the discharge air temperature increase after the engine has run for several minutes?
   Yes _____   No _____

9. Is there any steam from the vents or is there an odor of coolant in the discharge air?

   Yes _____   No _____

   Shut off the engine, and raise and support the vehicle to check the underside of the engine.

10. Inspect for signs of coolant loss from the water pump, lower radiator hose, and lower section of the radiator; note findings. _____

11. Summarize the results of your inspection. _____
    _____

12. Based on your findings, what are actions are necessary?
    _____
    _____

13. Remove any tools and equipment and clean up the work area.

    Instructor's check _____

Lab Worksheet 26-2

Name _____ Station _____ Date _____

Pressure Test Cooling System

ASE Education Foundation Correlation

This lab worksheet addresses the following **MLR** task:

1.C.1. Perform cooling system pressure and dye tests to identify leaks; check coolant condition and level; inspect and test radiator, pressure cap, coolant recovery tank, and heater core and gallery plugs; determine necessary action. **(P-1)**

Description of Vehicle

Year _____ Make _____ Model _____

Engine _____ AT MT CVT PSD (circle which applies)

Procedure

1. Determine the correct coolant for the vehicle using the owner's manual or manufacturer's service information.

   Source of information _____

   Recommended coolant type _____

2. Locate and record the cooling system operating pressure. _____

   Instructor's check _____

3. With the engine cool, carefully remove the radiator cap.

   Warning: Never open a hot cooling system. You can be severely burned by releasing hot coolant.

   Inspect and record coolant level and appearance.

   _____

   _____

4. Using the appropriate adapters, install the cooling system tester onto the radiator or pressure tank. Increase pressure to the limit shown on the radiator cap.

   Instructor's check _____

5. With the system under pressure, look for signs of coolant loss from the radiator, hoses, water pump, and other locations. Record your findings. _____

_____

6. Once you have inspected the system, slowly relieve the pressure and remove the pressure tester. Clean any spilled coolant and clean the pressure tester and adapters.

Instructor's check _____

7. Remove any tools and equipment and clean up the work area.

Instructor's check _____

Lab Worksheet 26-3

Name _____ Station _____ Date _____

Inspect and Replace Thermostat

ASE Education Foundation Correlation

This lab worksheet addresses the following **MLR** task:

1.C.3. Remove, inspect, and replace thermostat and gasket/seal. **(P-1)**

Description of Vehicle

Year _____ Make _____ Model _____

Engine _____ AT MT CVT PSD (circle which applies)

Procedure

1. Refer to the manufacturer's service information for the procedures to replace the thermostat. Summarize the steps. _____
_____

2. With the cooling system drained, remove the radiator hose and thermostat housing.
Instructor's check _____

3. Remove and inspect the thermostat. Is the thermostat stuck open?   Yes _____   No _____

4. Inspect the thermostat housing and water outlet. Note any problems. _____
_____

5. Using the service information, determine the thermostat's opening and full-open temperatures and normal engine operating temperature.

Thermostat begins to open at _____°F

Thermostat fully open at _____°F

Normal operating temperature _____°F

6. Install the new thermostat and seal or gasket. Install the bolts that hold the thermostat housing or water outlet in place. Housing bolt torque specifications. _____
Instructor's check _____

7. Torque the housing bolts to specifications and install the radiator hose.

   Instructor's check _____

8. Refill the cooling system with the correct type of coolant. Install the cooling system cap.

   Instructor's check _____

9. Start then engine and turn on the interior heat to the full Hot setting and turn on the fan. Connect a scan tool to the data link connector and navigate to the engine data menu. Locate the engine coolant temperature (ETC) data.

   Instructor's check _____

10. Start the engine and allow it to run until thermostat opening temperature is reached. Keep the coolant level full in the radiator or pressure tank.

    Instructor's check _____

11. Carefully measure the temperature of the upper and lower radiator hoses with an infrared thermometer or temperature probe for a DMM. Note the temperatures and compare them to what is displayed on the scan tool.

    Upper hose temp. _____°F

    Lower hose temp. _____°F

    ECT sensor temp. _____°F

12. If the thermostat is working properly, continue to let the engine run until the cooling fan turns on. Note the temperature displayed by the ECT sensor. If the fan is belt driven, record the highest temperature indicated by the ECT sensor.

    Coolant temp. when fan turn on _____

    Maximum coolant temp. (belt-driven fan) _____

13. Let the engine run until the cooling fan turns off. Note the temperature.

    Coolant temp. when fan turns off _____

    Minimum temp. (belt-driven fan) _____

14. Once you have confirmed that the thermostat is working properly, the cooling system is free of any trapped air, and the fan is operating correctly, clean up your work area and put away all tools and equipment.

    Instructor's check _____

Lab Worksheet 26-4

Name _____ Station _____ Date _____

Identify and Test Coolants

ASE Education Foundation Correlation

This lab worksheet addresses the following **MLR** task:

1.C.4. Inspect and test coolant; drain and recover coolant; flush and refill cooling system; use proper fluid type per manufacturer specification; bleed air as required. **(P-1)**

Procedure

Using a selection of vehicles, determine the correct coolant for each. To determine the correct coolant, check the owner's manual or service information. Check the condition of the coolant using either a coolant hydrometer or coolant test strip.

1. Year _____ Make _____ Model _____

    Engine _____

    Recommended coolant _____

    Cooling system capacity _____

    Note coolant appearance (note coolant level, color, and signs of contamination). _____

    _____

    Coolant protection temperature/pH balance _____

2. Year _____ Make _____ Model _____

    Engine _____

    Recommended coolant _____

    Cooling system capacity _____

    Note coolant appearance (note coolant level, color, and signs of contamination). _____

    _____

    Coolant protection temperature/pH balance _____

    _____

3. Year _____ Make _____ Model _____

    Engine _____

    Recommended coolant _____

    Cooling system capacity _____

    Note coolant appearance (note coolant level, color, and signs of contamination). _____
    _____

    Coolant protection temperature/pH balance _____

4. Based on your inspection, what can you determine about the condition of the coolant of each

    vehicle inspected? _____
    _____

    Instructor's check _____

Lab Worksheet 26-5

Name _____ Station _____ Date _____

Coolant Drain and Flush

ASE Education Foundation Correlation

This lab worksheet addresses the following **MLR** task:

1.C.4. Inspect and test coolant; drain and recover coolant; flush and refill cooling system with recommended coolant; bleed air as required. **(P-1)**

Description of Vehicle

Year _____ Make _____ Model _____

Engine _____ AT MT CVT PSD (circle which applies)

Procedure

1. Refer to the manufacturer's service information for specific procedures to drain, flush, refill, and bleed the cooling system. Summarize the steps. _____
    _____
    _____

    Steps 2–5 are for using a coolant service machine. Resume at Step 11.

2. If using a coolant recovery, flush and refill machine, connect the machine following the manufacturer's procedures. Describe how to connect the machine to the vehicle. _____
    _____
    _____

3. Determine the correct coolant for the vehicle and fill the coolant service machine with the specified coolant.

    Coolant type _____

    Instructor's check _____

4. Start the coolant recovery and refill process. Ensure the system is completely full when the cycle finished.

    Instructor's check _____

5. Disconnect the coolant service machine and top off the coolant as needed.

Instructor's check _____

Steps 6–10 are for those not using a coolant service machine.

6. If you are *not* using a coolant service machine, ensure the engine is cool and remove the cooling system cap. Note: Never open the cap on a hot cooling system as severe burns are possible. Next, locate the cooling system drain(s) and drain the coolant into the appropriate coolant waste tank or shop pans.

Instructor's check _____

7. Once as much of the old coolant has been drained as possible, use a flush gun or similar tool to flush the cooling system. Allow any remaining water to drain from the system and close the coolant drain(s).

8. Ensure you have the correct coolant as specified by the vehicle manufacturer.

Coolant type _____

Instructor's check _____

9. If a vacuum system is available, connect the tool to the cooling system fill neck and pull a vacuum on the system. Once under approximately 25 inch Hg of vacuum, open the valve to draw in the new coolant.

Instructor's check _____

10. If a vacuum system is not available, carefully pour new coolant into the system. Continue until the system is full.

Instructor's check _____

11. Once the system is full, start the engine and turn the heater to its full-hot setting. Let the engine run to allow any trapped air to escape from the cooling system. What concerns can be caused by air trapped in the cooling system? _____

_____

_____

12. Once the thermostat has opened and all trapped air has escaped, install the cooling system cap. Let the engine run until the radiator fan(s) cycle on and off.

Instructor's check _____

13. Remove any tools and equipment and clean up the work area.

Instructor's check _____

Lab Worksheet 26-6

Name _____ Station _____ Date _____

Identify Lubrication and Cooling System Components

ASE Education Foundation Correlation

This lab worksheet addresses the following **MLR** task:

1.C.6. Identify components of the lubrication and cooling systems. **(P-1)**

Description of Vehicle

Year _____ Make _____ Model _____

Engine _____ AT MT CVT PSD (circle which applies)

Procedure

1. Label the parts of the lubrication system shown here.

   a. _____

   b. _____

   c. _____

   d. _____

   e. _____

   f. _____

   g. _____

2. Using the service information, locate an exploded view or part explosion of the cooling system.

   Instructor's check _____

3. Locate the following parts of the cooling system on a vehicle.

a) Pressure tank/reservoir _____   g) Heater control valve _____

b) Radiator fan(s) _____   h) Heater core _____

c) Radiator _____   i) Heater hoses _____

d) Water pump _____   j) Thermostat _____

e) Radiator hoses _____   k) Air bleed valve _____

f) Coolant temperature sensor _____

Instructor's check _____

Lab Worksheet 26-7

Name _____ Station _____ Date _____

Research and TSB

ASE Education Foundation Correlation

This lab worksheet addresses the following **MLR** task:

7.A.1. Research vehicle service information, including refrigerant/oil type, vehicle service history, service precautions, and technical service bulletins. **(P-1)**

Description of Vehicle

Year _____ Make _____ Model _____

Engine _____ AT MT CVT PSD (circle which applies)

Procedure

1. Determine the refrigerant type and charge amount for this vehicle. _____

   Refrigerant _____

   Charge amount _____

2. Determine the refrigerant oil type and charge amount.

   Refrigerant oil _____

   Capacity _____

3. Locate and note any service precautions for working on the air conditioning system for this vehicle.

   _____

   _____

   _____

4. Using the service information, locate a technical service bulletin (TSB) for the HVAC system.

   TSB number, date, and topic: _____

   _____

   _____

   Instructor's check _____

Lab Worksheet 26-8

Name _____ Station _____ Date _____

ID Components

ASE Education Foundation Correlation

This lab worksheet addresses the following **MLR** task:

7.A.2. Identify heating, ventilation and air conditioning (HVAC) components and configuration. **(P-1)**

Description of Vehicle

Year _____ Make _____ Model _____

Engine _____ AT MT CVT PSD (circle which applies)

Procedure

1. Identify the major A/C system components shown in Figure 1.

   1. _____

   2. _____

   3. _____

   4. _____

   5. _____

   6. _____

   7. _____

2. On a vehicle. Locate the following:

a. A/C system label

b. Low-side service port

c. High-side service port

d. Dryer and accumulator

e. Compressor

f. Metering device

g. Condenser

h. High- and low-pressure switch

Instructor's check _____

Lab Worksheet 26-9

Name _____ Station _____ Date _____

A/C Compressor Belt

ASE Education Foundation Correlation

This lab worksheet addresses the following **MLR** task:

7.B.1. Inspect and replace A/C compressor drive belts, pulleys, and tensioners; visually inspect A/C components for signs of leaks; determine necessary action. **(P-1)**

Description of Vehicle

Year _____ Make _____ Model _____

Engine _____ AT MT CVT PSD (circle which applies)

Procedure

1. Locate and note the type of drive belt used.

   V-belt _____   Serpentine (multirib) belt _____

2. Describe how tension is applied to the belt. _____

   _____

3. Inspect the drive belt for wear and damage. Note your findings. _____

   _____

   _____

4. Locate the belt tension specifications and record them.

   Specification _____

5. Using a belt tension gauge, measure and record the generator drive belt tension.

   Measured tension _____

6. How does the measured tension compare to the specification? _____
_____

7. Based on your inspection, what do you recommend? _____
_____

_____

8. Remove any tools and equipment and clean up the work area.

Instructor's check _____

Lab Worksheet 26-10

Name _____ Station _____ Date _____

Inspect A/C Components

ASE Education Foundation Correlation

This lab worksheet addresses the following **MLR** task:

7.B.1. Inspect and replace A/C compressor drive belts, pulleys, and tensioners; visually inspect A/C components for signs of leaks; determine necessary action. **(P-1)**

Year _____ Make _____ Model _____

Engine _____ AT MT CVT PSD (circle which applies)

1. Perform a visual inspection of the A/C system. Note any faults. or concerns with the following:

   a. A/C compressor drive belt   OK _____   Not OK _____

   b. A/C compressor   OK _____   Not OK _____

   c. A/C hoses   OK _____   Not OK _____

   d. Cooling fans   OK _____   Not OK _____

2. Visually inspect the A/C components for signs of refrigerant loss. Note your findings. _____
   _____
   _____

3. Close all but the center vent on the dash and place a thermometer into the open dash vent. Note the temperature inside the vehicle before the A/C is turned on.

   Temperature _____°F

4. Start the engine, set the air to discharge from the dash vents, place the A/C on MAX cooling, and set the blower on high speed. Run the engine at 2,000 rpm and note the temperature of the air from the vent. Run the system for a couple of minutes until temperature remains steady.

   Temperature _____°F

5. Shut the A/C off and then shut the engine off. Using the service information, locate the A/C performance test temperature chart. Using the chart, determine the condition of the A/C system and note your results.

   _____

6. Remove any tools and equipment and clean up the work area.

   Instructor's check _____

Lab Worksheet 26-11

Name _____ Station _____ Date _____

ID HEV A/C

ASE Education Foundation Correlation

This lab worksheet addresses the following **MLR** task:

7.B.2. Identify hybrid vehicle A/C system electrical circuits and the service/safety precautions. **(P-2)**

Description of Vehicle

Year _____ Make _____ Model _____

Engine _____ AT MT CVT PSD (circle which applies)

Procedure

1. Using the service information, locate and summarize any safety precautions for working on or near the air conditioning system for this vehicle. _____

   _____

   _____

   _____

2. Using the service information, locate and summarize any service precautions for working on the air conditioning system for this vehicle. _____

   _____

   _____

3. What color is the high-voltage wiring in this vehicle? _____

4. Summarize the procedures to safely disconnect the high-voltage system on this vehicle.

   _____

   _____

   _____

5. Visually locate any high-voltage wiring for the air conditioning system and note the locations.

   _____

   _____

   _____

Lab Worksheet 26-12

Name _____ Station _____ Date _____

Condenser Airflow

ASE Education Foundation Correlation

This lab worksheet addresses the following **MLR** task:

7.B.3. Inspect A/C condenser for airflow restrictions; determine necessary action. **(P-1)**

Description of Vehicle

Year _____ Make _____ Model _____

Engine _____ AT MT CVT PSD (circle which applies)

Procedure

1. Describe what concerns can be caused by airflow across the condenser being restricted. _____
   _____
   _____
   _____

2. Describe two possible causes of airflow restrictions across the condenser. _____
   _____
   _____
   _____

3. Inspect the vehicle for any signs of restricted airflow around the condenser. Note your findings.
   _____
   _____
   _____

4. Based on your inspection, what is the necessary action? _____
   _____
   _____

5. Remove any tools and equipment and clean up the work area.

   Instructor's check _____

Lab Worksheet 26-13

Name _____ Station _____ Date _____

Inspect Cooling System Hoses

ASE Education Foundation Correlation

This lab worksheet addresses the following **MLR** task:

7.C.1. Inspect engine cooling and heater systems hoses and pipes; determine necessary action. **(P-1)**

Description of Vehicle

Year _____ Make _____ Model _____

Engine _____ AT MT CVT PSD (circle which applies)

Procedure

Inspect the following components and note your findings.

1. Pressure cap rating _____ psi   Bar   kPa (circle one)

2. Check for leaks around pressure cap. Note your findings. _____

3. Inspect the radiator for signs of leaks. Note your findings. _____

4. Inspect hoses for leaks, cracks or bulges. Note your findings. _____

5. Inspect radiator fan for damage. Note your findings. _____

6. Inspect for coolant leaks around the thermostat and hoses. Note your findings. _____

   _____

   Start the engine and set the heater controls to high heat output.

7. Does the discharge air temperature increase after the engine has run for several minutes?

   Yes _____   No _____

8. Is there any steam from the vents or is there an odor of coolant in the discharge air?

   Yes _____   No _____

   Shut off the engine, and raise and support the vehicle to check the underside of the engine.

9. Inspect for signs of coolant loss from the water pump, lower radiator hose, and lower section of the radiator; note findings. _____

10. Summarize the results of your inspection. _____

_____

_____

11. Based on your findings, what are actions are necessary? _____

_____

_____

12. Remove any tools and equipment and clean up the work area.

Instructor's check _____

Lab Worksheet 26-14

Name _____ Station _____ Date _____

Inspect A/C Ducts

ASE Education Foundation Correlation

This lab worksheet addresses the following **MLR** task:

7.D.1. Inspect A/C-heater ducts, doors, hoses, cabin filters, and outlets; determine necessary action. **(P-1)**

Description of Vehicle

Year _____ Make _____ Model _____

Engine _____ AT MT CVT PSD (circle which applies)

Procedure

1. Refer to the manufacturer's service information for specific inspection procedures. Summarize the procedures. _____

    _____

    _____

2. What types of concerns can be caused by problems with the A/C-heater ducts, doors, hoses, cabin filters, or outlets? _____

    _____

    _____

3. Visually inspect all A/C-heater ducts, doors, hoses, filters, and outlets that are accessible. Note your findings. _____

    _____

    _____

    Instructor's check _____

4. With the ignition on, operate the A/C-heater system and check the operation of each of the following items:

    Airflow from defrost vents   Yes _____   No _____

    Airflow from panel vents   Yes _____   No _____

    Airflow from floor vents   Yes _____   No _____

    Airflow temperature changes with setting changes Yes _____   No _____

5. For vehicles with dual- or multi-zone systems, check for airflow and temperature changes for each zone. Note your findings. _____

_____

6. Locate, remove, and inspect the cabin air filter. Note your findings. _____

_____

_____

7. Based on your inspection, what are the necessary actions? _____

_____

_____

8. Remove your tools and equipment and clean up the work area.

Instructor's check _____

Lab Worksheet 26-15

Name _____ Station _____ Date _____

ID A/C Odor

ASE Education Foundation Correlation

This lab worksheet addresses the following **MLR** task:

7.D.2. Identify the source of A/C system odors. **(P-2)**

Description of Vehicle

Year _____ Make _____ Model _____

Engine _____ AT MT CVT PSD (circle which applies)

Procedure

1. Refer to the manufacturer's service information for specific odor inspection procedures. Summarize the procedures. _____

   _____

   _____

   _____

2. What types of concerns can cause odors from the A/C system? _____

   _____

   _____

   _____

3. Visually inspect all A/C-heater ducts, doors, hoses, filters, and outlets that are accessible. Note your findings. _____

   _____

   _____

   Instructor's check _____

4. If no problems are found during your inspection, what is the likely source of the odor from the A/C system?

   _____

   _____

5. How can this problem be confirmed? _____

_____

6. How can this problem be corrected? _____

_____

7. Remove any tools and equipment and clean up the work area.

Instructor's check _____

Lab Worksheet 26-16

Name _____ Station _____ Date _____

Verify Engine Temperature

ASE Education Foundation Correlation

This lab worksheet addresses the following **MLR** task:

8.A.6. Verify engine operating temperature. **(P-1)**

Description of Vehicle

Year _____ Make _____ Model _____

Engine _____ AT MT CVT PSD (circle which applies)

Procedure

1. Using the service information, determine thermostat opening and fully open temperatures and normal engine operating temperature.

   Thermostat begins to open at _____°F

   Thermostat fully open at _____°F

   Normal operating temperature _____°F

2. Connect a scan tool to the data link connector and navigate to the engine data menu. Locate the engine coolant temperature (ETC) data. Note starting temperature.

   Starting temperature _____°F(C)

3. Start the engine and allow it to run until thermostat opening temperature is reached. Next, carefully measure the temperature of the upper and lower radiator hoses with an infrared thermometer or temperature probe for a DMM. Note the temperatures and compare them to what is displayed on the scan tool.

   Upper hose temperature _____°F

   Lower hose temperature _____°F

   ECT sensor temperature _____°F

4. Based on the temperatures, is the thermostat working properly?

   Yes _____   No _____

5. If the temperatures are too low, what does this indicate? _____

_____

6. If the ECT sensor temperature keeps increasing beyond thermostat opening temperature but the temperature of the radiator hoses is much cooler, what can this indicate? _____

_____

_____

_____

7. If the thermostat is working properly, continue to let the engine run until the cooling fan turns on. Note the temperature displayed by the ECT sensor. If the fan is belt driven, record the highest temperature indicated by the ECT sensor.

Coolant temperature when fan turn on _____

Maximum coolant temperature (belt-driven fan) _____

8. Let the engine run until the cooling fan turns off. Note the temperature.

Coolant temperature—when the fan turns off _____

Minimum temperature (belt-driven fan) _____

9. Based on your inspection, what is the condition of the thermostat and cooling system operation?

_____

_____

_____

10. Remove any tools and equipment and clean up the work area.

Instructor's check _____

CHAPTER 27

Vehicle Maintenance

Review Questions

1. _____ is the act of keeping something in a state of good operating condition.

2. Maintenance is performed to prolong the life of something, and a repair is made to correct a _____.

3. Using the list provided in the textbook, make a list of maintenance items you already have performed on your own vehicle or someone else's vehicle. _____
   _____
   _____
   _____
   _____
   _____

4. Periodic inspection and service to keep the vehicle operating in good condition provides _____ for the technician and the shop.

5. Describe how to use maintenance as a way to build a relationship with your customers.
   _____
   _____
   _____
   _____

6. Failure to perform routine maintenance can cause serious _____ to the vehicle.

7. Maintenance information is usually located in the vehicle's _____.

8. Describe the differences between what may be considered normal service and severe service. _____
   _____
   _____

9. Once a periodic maintenance service has been performed, it is important to _____ the maintenance reminder or timer.

10. Why is it important to review the customer's service history before recommending some services be performed? _____

_____

_____

11. Before moving any vehicle for service, always check brake pedal _____ and _____ before you begin to move the vehicle.

12. Why should a check of TSBs be performed as part of determining what maintenance is necessary? _____

_____

_____

13. While tire pressure is being checked, *Technician A* says higher-than-specified tire pressure may indicate the tires have been recently driven on. *Technician B* says if the tire pressure is above the specified pressure, the pressure should be reduced. Who is correct?
    a. Technician A
    b. Technician B
    c. Both A and B
    d. Neither A nor B

14. Describe how to quickly check shock absorber condition. _____

_____

_____

15. Why must you use caution when replacing wiper blades? _____

_____

_____

_____

16. If _____ is found in the housing, perform a thorough inspection of the PCV system.

17. Describe three common drive belt wear conditions. _____

_____

_____

_____

18. _____ help filter out pollen, dirt, and odors from the air entering the passenger compartment.

19. Explain the steps of performing a chassis lubrication. _____
_____
_____
_____

20. List what lights should be checked as part of the maintenance inspection. _____
_____
_____
_____

21. List five fluids that should be checked during a maintenance inspection.
    a. _____
    b. _____
    c. _____
    d. _____
    e. _____

22. List four types of coolants in use in modern vehicles._____
_____
_____

23. *Technician A* says green coolant is universal and can be used in any type of vehicle. *Technician B* says color has little to do with choosing the correct coolant. Who is correct?
    a. Technician A
    b. Technician B
    c. Both A and B
    d. Neither A nor B

24. Which of the following statements about coolant is correct?
    a. Some coolants contain silicates.
    b. Some coolants have no silicates.
    c. Silicates are added to coolant to prevent corrosion.
    d. None of the above.

25. New coolant should be mixed with _____ water.

26. *Technician A* says all automatic transmissions have a dipstick for checking fluid level. *Technician B* says most transmissions can use any type of automatic transmission fluid. Who is correct?
    a. Technician A
    b. Technician B
    c. Both A and B
    d. Neither A nor B

27. Diesel _____ fluid is used to help reduce exhaust emissions and requires periodic refilling.

28. DEF is composed of 33 percent _____ and 67 percent pure _____.

29. Explain why exhaust leaks can be dangerous. _____
_____
_____

30. Explain viscosity as it applies to engine oil. _____
_____
_____
_____

31. What does multiviscosity oil mean? _____
_____
_____

32. Which of the following engine oil ratings are important to understand?
    a. SAE and API
    b. ILSAC and ACEA
    c. Manufacturer-specific oil ratings
    d. All of the above

33. Which of the following are problems that may result from using the incorrect engine oil?
    a. Internal engine sludge
    b. Variable valve timing faults
    c. Engine failure
    d. All of the above

34. Before beginning an oil change, take time to check and document the engine oil _____ and
    _____.

35. After removing the old oil filter, clean and _____ the filter gasket surface on the engine.

36. Most modern gasoline engines hold approximately how much engine oil?
    a. Two to four quarts
    b. Four to seven quarts
    c. One to two gallons
    d. Six to ten quarts

Activities

1. Organize the following items into two categories: those that are typically replaced as maintenance and those replaced as part of a repair.

Water pump	Timing belt
Generator	Fuel pump
Air filter	Wiper blades
Brake pads	Wheel bearing
Spark plugs	Ignition coil
Wheel speed sensor	Tie rod end
Cabin air filter	Fuel filter
Accessory drive belt	Power steering pump

2. Organize the following fluids into two categories: applications that often require specific fluid types and qualities and applications that do not.

Brake fluid	Automatic transmission fluid
Engine oil	Power steering fluid
Windshield washer fluid	Differential lubricant
Antifreeze	Hydraulic clutch fluid

   a. How can a technician determine if specific fluids are required for an application?

   _____

   _____

   _____

   b. Explain why it is important to use the correct fluid as specified by the vehicle manufacturer. _____

   _____

   _____

Lab Activity 27-1

Name _____ Date _____ Instructor _____

Year _____ Make _____ Model _____

Engine _____   AT   MT (circle one)

Maintenance Schedules

Locate the manufacturer's maintenance schedules for this vehicle.

1. How does the manufacturer define severe service? _____
_____
_____

2. Based on your experiences, how likely do you think it is that the vehicle is operated under severe conditions? Explain your answer. _____
_____

3. How does the severe schedule differ from the normal service schedule? _____
_____
_____

4. Is the vehicle equipped with a towing package or trailer hitch?

Yes _____   No _____

If yes, how can pulling a trailer affect the maintenance requirements for the vehicle? _____
_____
_____

Lab Activity 27-2

Name _____ Date _____ Instructor _____

Year _____ Make _____ Model _____

Engine _____   AT   MT (circle one)

Fluid Inspection

1. Locate the manufacturer's recommended fluid types for the following:

   Engine oil   _____

   Transmission   _____

   Brake fluid   _____

   Power steering   _____

   Antifreeze   _____

   Differential   _____

2. Check each of the following fluids (note the fluid level and color).

   Engine oil   _____

   Transmission   _____

   Brake fluid   _____

   Power steering   _____

   Antifreeze   _____

   Differential   _____

3. Based on your inspection, what actions should be performed? _____

   _____

   _____

   _____

Lab Worksheet 27-1

Name _____ Station _____ Date _____

Inspect IP Indicators

ASE Education Foundation Correlation

This lab worksheet addresses the following **MLR** task:

1.A.2.   Verify operation of the instrument panel engine warning indicators. **(P-1)**

Description of Vehicle

Year _____ Make _____ Model _____

Engine _____ AT MT CVT PSD (circle which applies)

Procedure

1. When the ignition is on, the instrument panel should turn on all indicator lights and perform a self-test. Turn the ignition to ON or RUN and note that the following light up:

   On   Off   Check engine light (MIL)

   On   Off   Battery or charging system light

   On   Off   Oil pressure light

   On   Off   Electronic throttle control light

   On   Off   Coolant temperature light

2. With the ignition still on, note which lights remain on after 15 seconds. _____

   _____

3. Start the engine and note the lights. Do any lights remain on?   Yes _____   No _____

4. What does it mean if a light remains on with the engine running? _____

   _____

5. With the engine running, note the readings of the coolant temperature, oil pressure, and charging system gauges.

Coolant temperature _____

Oil pressure _____

Charging system _____

6. Based on your observations, describe the operating condition of the engine and its subsystems._____

_____

_____

Lab Worksheet 27-2

Name _____ Station _____ Date _____

Engine Oil Change

ASE Education Foundation Correlation

This lab worksheet addresses the following **MLR** task:

1.C.5. Perform engine oil and filter change; use proper fluid type per manufacturer specification; reset maintenance reminder as required. **(P-1)**

Description of Vehicle

Year _____ Make _____ Model _____

Engine _____ AT MT CVT PSD (circle which applies)

Procedure

1. Before starting to change the engine oil, refer to the service information to determine the oil type and capacity for the vehicle.

   Oil rating specifications _____

   Oil capacity _____

2. Why is it important that the correct oil type be used for each vehicle? _____

   _____

3. Open the hood and install fender covers to protect the vehicle.

   Instructor's check _____

4. If applicable, remove the oil dipstick and note oil level and appearance. _____

   _____

5. Remove the oil fill cap and inspect the underside of the cap for signs of moisture or oil sludge. Note the condition of the oil fill cap and set the cap aside in a safe location. _____

   _____

6. If the vehicle has a canister-type oil filter: loosen the filter housing and remove the filter. Place the filter and housing in the oil drain pan until you are ready to install the new filter.

7. Raise the vehicle and place a drain pan under the oil pan drain plug. Remove the plug and drain the oil. Locate the oil drain plug torque specification and record it here.

   Drain plug torque specification _____

8. Inspect the oil drain plug and washer. Once the oil has drained sufficiently, reinstall the drain plug and tighten to specifications.

   Instructor's check _____

9. If the vehicle has a cartridge oil filter: Place the pan under the oil filter and remove the oil filter. Clean the filter mounting base and ensure the filter gasket is not stuck to the filter mounting base.

   Instructor's check _____

10. Prefill the oil filter with new oil if applicable. Place a few drops of new oil on the filter gasket and lubricate the gasket. Carefully thread the new filter onto the mounting base and tighten per the filter manufacturer's directions.

    Instructor's check _____

11. Lower the vehicle. If installing a canister filter: verify the new filter matches the old filter. Install the new O-ring(s) onto the housing. Install the new filter and housing. Torque the housing to specification.

    Filter housing torque specification _____

    Instructor's check _____

12. Using a funnel, begin pouring the oil into the engine. Do not overfill the engine—refer to the capacity specifications as needed. Remove the funnel and clean any spilled oil from the engine compartment.

    Instructor's check _____

13. Install the oil filler cap. With clean hands, start the engine. Note the oil pressure light or gauge on the dash. The oil light should go out after a few seconds of engine running. The oil pressure gauge should show oil pressure after a few seconds of engine running. Check under the vehicle for signs of oil leaks. Shut the engine off.

    Instructor's check _____

14. Recheck the oil level. Add oil as necessary to bring the oil level to the full mark on the dipstick.

    Instructor's check _____

15. Determine how to reset the oil life index or maintenance reminder. Summarize the procedure here.

    _____

    _____

16. Reset and verify the oil life index. Remove the fender covers and perform a final check under the hood. Ensure the oil filler cap is tight and that no oil has been spilled under the hood.

    Instructor's check _____

17. Once finished, put away all tools and equipment and clean up the work area.

    Instructor's check _____

Lab Worksheet 27-3

Name _____ Station _____ Date _____

Inspect Wiper/Washer

ASE Education Foundation Correlation

This lab worksheet addresses the following **MLR** task:

6.E.8. Verify windshield wiper and washer operation; replace wiper blades. **(P-1)**

Description of Vehicle

Year _____ Make _____ Model _____

Wiper washer inspected _____ Front _____ Rear _____

Procedure

1. Perform a visual inspection of the windshield wiper blades. Note your findings.

_____

2. Turn the ignition on and activate the front windshield washer.

Washer pump operation	OK _____	Not OK _____
Washer spray amount	OK _____	Not OK _____
Washer spray pattern	OK _____	Not OK _____
Wiper effectiveness	OK _____	Not OK _____
Wiper noise	OK _____	Not OK _____

3. Turn the ignition on and activate the rear windshield washer.

Washer pump operation	OK _____	Not OK _____
Washer spray amount	OK _____	Not OK _____
Washer spray pattern	OK _____	Not OK _____
Wiper effectiveness	OK _____	Not OK _____
Wiper noise	OK _____	Not OK _____

4. Based on your inspection, what is the condition of the wipers and washer? _____

_____

_____

5. Before replacing the wiper blades, place a clean fender cover over the fender and across the windshield under the wipers. This is to help prevent damage to the windshield if the wiper arm contacts the glass.

         Instructor's check _____

6. To replace the wiper blades, inspect the blade and wiper arm and determine how the blade is attached to the wiper arm. Release the blade from the arm and remove the blade.

         Instructor's check _____

7. Compare the replacement wiper blade to the old blade and make sure it is correct for the vehicle.

         Instructor's check _____

8. Install the new wiper blade(s). Make sure the blades fully lock onto the wiper arm. Remove the fender cover from the windshield and carefully set wiper blades onto the glass.

         Instructor's check _____

9. Turn the ignition on and activate the wiper and washer system. Make sure the wipers fully contact the windshield and remove the washer fluid.

         Instructor's check _____

10. Remove your tools and equipment and clean up the work area.

         Instructor's check _____

Lab Worksheet 27-4

Name _____ Station _____ Date _____

Inspect DEF

ASE Education Foundation Correlation

This lab worksheet addresses the following **MLR** task:

8.C.5. Check and refill diesel exhaust fluid (DEF). **(P-2)**

Description of Vehicle

Year _____ Make _____ Model _____

Engine _____ AT MT CVT PSD (circle which applies)

Procedure

1. Refer to the manufacturer's service information for specific procedures to inspect and fill the diesel exhaust fluid (DEF) system. Summarize the procedures. _____

   _____

   _____

2. Determine DEF level.

   Good _____ Marginal _____ Low _____

3. What can result if the DEF tank is left empty and the vehicle continues to be driven?

   _____

   _____

   _____

4. Note any special safety precautions for working with DEF. _____

   _____

5. Refill the DEF tank as necessary.

   Instructor's check _____

6. Remove any tools and equipment and clean up the work area.

   Instructor's check _____